Quantum Geometry in Diamond Defects

Wesley A. Richardson

Abstract

Recent years have witnessed rapid development in the field of quantum information science. Quantum technologies are promising to revolutionize many fields, ranging from fast computation and simulation to precise sensing and secure communication. While the performance of quantum devices in the current noisy intermediate-scale quantum (NISQ) era is still limited, exploring intriguing applications of various quantum systems has been an active research area. Among the many physical platforms, solid-state spin defects, such as nitrogen-vacancy (NV) center in diamond, have attracted a lot of attentions thanks to their good controllability and long coherence time even at room temperature.

In this thesis, we present our efforts in demonstrating promising quantum applications based on the NV system. With microwave and optical pulses, we have good control on the NV center system, which enables us to engineer the desired Hamiltonian and probe the information of quantum states. In particular, the geometry of a quantum state is characterized by the quantum geometric tensor (QGT), which can find applications in simulating topological material and quantum metrology. We experimentally measure the QGT of an engineered Hamiltonian in a single NV center using weak periodic modulation method. Based on it, we are able to reveal the existence of a tensor monopole that characterizes a tensor gauge field in parameter space. Furthermore, we find that the QGT plays a significant role in quantum multi-parameter estimation. It not only quantifies the precision limit when estimating parameters but also links to the attainability of precision bounds.

Thanks to the sensitivity studied above, NV centers have emerged as powerful quantum sensors to detect various signals, ranging from electromagnetic fields to temperature, with high sensitivity and spatial resolution. Detecting biological or chemical signals is more challenging, but a promising avenue is to transduce them into magnetic noise that the NV center is sensitive to. We demonstrate this strategy by measuring the rotational Brownian motion of magnetic molecules. Then, exploiting the dipolar interaction between NV centers and magnetic molecules, we design a hybrid sensor that is capable of detecting the SARS-CoV-2 virus RNA with ultrahigh sensitivity and low false negative rate. Finally, we will show that the charge state of NV centers is sensitive to surface modification of nanodiamond, which enables us to design alkali ion sensors with the help of chemical engineering.

Contents

List of Figures

List of Tables

Chapter 1

Introduction

1.1 Quantum information science

At the most microscopic level, all physical systems are governed by quantum mechanics. Counter-intuitive effects resulting from quantum theory, such as superposition states and quantum entanglement, have enabled many significant applications that are out of reach of corresponding classical devices. Recent years have witnessed dramatic development in quantum technologies, including applications in the field of computation, communication, simulations and metrology with quantum mechanical systems. Quantum computers could perform the tasks of, for example, quantum search [8] or factoring large numbers [9] that are hard or impossible for even the most powerful classical computers; quantum communication [10, 11] could enable information transfer or teleportation between distant nodes with its security guaranteed by fundamental laws of quantum mechanics; quantum simulation [12], on the other hand, has the ability to simulate complex systems in the field of condensed-matter physics, particle physics or chemical reactions and drug designs; another significant application, quantum sensing [13], is among the most established applications of quantum information science and could in principle operate at greatest possible sensitivity and precision that are allowed by physical laws.

Implementation of these exciting applications in various quantum devices has been a very active research area. Different quantum systems, such as superconducting circuits, trapped ions and neutral atoms, have found their advantages in different fields of applications. Among the many physical quantum platforms, solid-state spin defects, such as nitrogen-vacancy (NV) center in diamond, have attracted a lot of attentions thanks to their good controllability and long coherence time even at room temperature. The NV system would be the focus of study throughout this thesis and we will present several aspects of developing different quantum technologies using NV centers.

1.2 Nitrogen-vacancy center in diamond

1.2.1 Introduction on NV centers

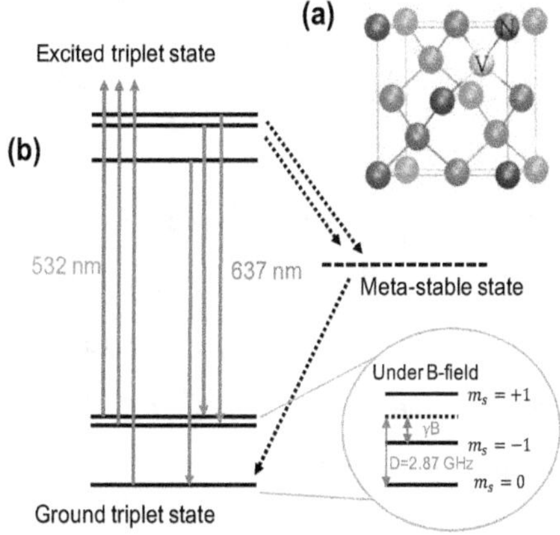

Figure 1-1: (a) Structure of NV centers in diamond. A single NV center consists of a nitrogen atom with a nearby vacancy. (b) Energy level diagram of the negatively charged NV center. A green laser will excite the spin to excite triplet states and then it will decay back with red fluorescence. The resonant fluorescence wavelength is 637 nm, known as the zero-phonon line (ZPL). Note that the $|m_s = \pm 1\rangle$ excited states can also decay back to $|m_s = 0\rangle$ ground state via a non-radiative process (dashed lines), hence a state initially at $|m_s = \pm 1\rangle$ will yield a smaller photon counts. An applied magnetic field can lift the degeneracy of ground states $|m_s = \pm 1\rangle$ (inset).

We start by giving a brief introduction on the NV system, including its energy levels and system control and characterization. The NV defect consists of a vacancy with an adjacent nitrogen atom in substitute of a carbon atom, as show in Fig. 1-1 (a). As a single defect in diamond, a NV center combines the advantage of atomic scale and robustness of solid-state systems. We focus on the negatively charged NV state unless stated. The energy levels of a single NV center consist of ground and excited spin triplet states and a metastable singlet state. The NV spin can be excited from the ground state to the excited state with a green laser (532 nm) and then it decays back while emitting red fluorescence. More importantly, the intersystem crossing (ISC) process shown in Fig. 1-1 (b) yields different fluorescence intensity, depending on whether the spin is in $|m_s = 0\rangle$ and $|m_s = \pm 1\rangle$ ground state initially. Hence on one hand, the spin state could be efficiently readout by detecting this fluorescence difference and on the other hand, a long-enough green laser pulse would efficiently polarize the spin into $|m_s = 0\rangle$ state. The property of high fidelity spin state readout and preparation in a fully optical way makes NV stand out in many applications. The readout fidelity might be further increased by exploiting machine learning methods [14].

Moreover, its ground state such as $|m_s = 0\rangle$ and $|m_s = -1\rangle$ (or $|m_s = +1\rangle$) could be utilized either as a qubit to encode information or as a sensor to measure external physical quantities. The energy splitting between these two levels can be tuned with an external magnetic field via the Zeeman effect (inset of Fig. 1-1 (b)). The ground state Hamiltonian of a single NV center in the presence of external magnetic field $\mathbf{B}$ is given by

$$H_{NV,g} = DS_z^2 + \gamma \mathbf{B} \cdot \mathbf{S}, \tag{1.1}$$

where $\mathbf{S} = (S_x, S_y, S_z)$ is the electronic spin-1 operator, $D = 2.87$ GHz is the zero-field splitting resulting from an electronic spin-spin interaction within the NV center and $\gamma = 2.8$ MHz/G is the gyromagnetic ratio. By applying a microwave field, one can create superposition states and perform logic operations between the two levels with a high fidelity. Specifically, noise filtering techniques such as dynamical decoupling [15, 16, 17, 18] using composite microwave pulses can be applied to decouple unwanted noise. The exquisite quantum control afforded by NV-centers makes them a good platform for engineering desired Hamiltonian and probing information of prepared quantum states.

If we consider all the three energy levels $|m_s = 0, \pm 1\rangle$ of the NV ground state in the presence of an external magnetic field, they form a qutrit system and we can apply dual-frequency microwave pulses that are on resonance with the $|m_s = 0\rangle \leftrightarrow |m_s = -1\rangle$ and $|m_s = 0\rangle \leftrightarrow |m_s = +1\rangle$ transition to control it. We will present a detailed discussions on the dual-frequency control in Chapter 2. Note that the $|m_s = +1\rangle \leftrightarrow |m_s = -1\rangle$ transition has a change of quantum number by two and cannot be directly controlled via a single microwave pulse. To achieve such direct control, phonon-based devices such as mechanical oscillators may be used.

We next discuss the relaxation mechanism for NV center. If prepared in the $|m_s = 0\rangle$ state, due to interaction with its environment, the spin state will finally decay to thermal equilibrium state with the spin state jumping from $|m_s = 0\rangle$ to $|m_s = \pm 1\rangle$ states. At room temperature, this thermal state corresponds to a fully mixed state. We can optically read out the spin state and extract the time scale of this process. This process is known as longitudinal relaxation and the related characteristic time is T_1. We will give a theoretical model of the T_1 relaxation process and discuss its applications in Chapter 4 and Chapter 5. A typical confocal setup to measure the T_1 time is depicted in Appendix. B. If prepared in a superposition state of $|m_s = 0\rangle$ and $|m_s = +1\rangle$ (or $|m_s = -1\rangle$) state, the loss of coherence can be characterized with a time scale T_2^*, known as dephasing process. By detecting the relaxation times or the electron spin resonance of NV centers, one can extract the information on local changes of magnetic field, temperature and so on. One of the most fascinating property of NV systems is that it can operate at room temperature while its ground state has long coherence time and the polarization of NV spin state does not need low temperature. In terms of quantum computation, the long coherence time of NV center and nearby nuclear spins such as ^{14}N (or ^{15}N) and ^{13}C makes the system a good candidate for quantum memory [19, 20, 21, 22, 23, 24, 25]. In quantum sensing protocols, this long coherence time would correspond to a long interrogation time thus enabling high sensitivity [13, 26].

1.2.2 NV-based quantum applications

In recent years, NV centers have been used for various quantum applications, including quantum computation, simulation, sensing and communication. In this section, we will present a brief overview of NV center's applications, particularly in quantum sensing and network.

Quantum sensing

Quantum sensing describes exploiting quantum coherence or entanglement to measure a physical quantify, such as magnetic field, temperature and strain, with a quantum object [13]. The use of counterintuitive quantum effects such as coherence [27], entanglement [28], squeezing [29, 30] or other nonclassical correlation [31, 32] can attain measurement accuracy and precision that go beyond the standard quantum limit. From precise quantum clocks [33], magnetometers [34] to gravitational wave detection such as LIGO [35], exciting applications of quantum sensing have arisen in recent years.

Experimental implementations of quantum sensors include atoms or trapped ions, solid-state spins, superconducting circuits and so on. In particular, NV centers in diamond act as stable fluorescence markers and magnetic field sensors, and have been investigated as quantum sensors for applications ranging from material science to chemistry and biology [26].

NV centers can respond to external signals through different mechanisms. For example, the energy levels of the NV center's ground state can be shifted by signals including magnetic field [26, 36, 37], temperature [38, 39, 40, 41], electrical field [42, 43, 44] and strain [45, 46, 38]. The hyperfine interaction strength between NV center and nearby nitrogen and carbon 13 nuclear spins is also shown to be temperature- and pressure-dependent [47, 48, 49, 50, 51, 52].

In particular, NV center has been used as magnetometers to detect DC or AC magnetic fields. In NV-based magnetometry, one is interested in the magnetic field's magnitude, direction, or projection onto a particular axis. Magnetometry measurements based on pulsed sequences can benefit from long sensing intervals τ. For example, in a typical Ramsey experiment, the accumulated magnetic-field-dependent phase $\phi = \gamma B \tau$ increases with time τ and maximal sensitivity to changes in B is achieved when τ is large. At the same time, due to inevitable interaction between NV center and its environment, signal contrast degrades with increasing interrogation time. Therefore, τ is limited by relaxation times T_2^*, T_2 and T_1, which characterize dephasing, decoherence, and spin-lattice interactions respectively. More specifically, DC sensitivity of NV-based magnetometry is limited to dephasing time T_2^*, while the AC sensitivity is limited by the coherence time T_2 that is usually one or two orders of magnitude longer than T_2^*. Note that T_2 can be prolonged through use of dynamical decoupling protocols [15, 16, 17] to approach the longitudinal spin relaxation time T_1. The sensitivity of the magnetic field sensors can thus be optimized by extending the relaxation times via novel driving sequences and material engineering [34].

In terms of detection frequency range, DC sensing protocols are targeted at static, slowly-varying, or broadband near-DC signals [53], whereas AC sensing protocols usually focus on narrowband, time-varying signals at frequencies up to around 1 MHz to 100 MHz [54, 55, 34, 56, 57]. Recently, sensing of arbitrary-frequency magnetic fields is achieved using a quantum mixer [58]. By leveraging nonlinear effects in periodically driven (Floquet) NV system [59, 60, 61], quantum frequency mixing of the signal and an applied bias AC field was demonstrated.

Alternatively, T_1 relaxometry allows phase-insensitive detection of signals at frequencies in the GHz regime and is T_1-limited. NV-based T_1 relaxometry [62, 63] has witnessed rapid development recently and the technique has been deployed in sensing of various signals, including biological processes and chemical events [64, 65, 66, 67, 68, 69, 70, 71, 72]. We will leave the detailed discussions of its applications in Chapter 4 and Chapter 5.

While the above protocols have focused on the negatively charged NV center, NV centers are also present in the neutral charge state NV^0 (spin 1/2). NV^- can lose an electron and becomes NV^0 state and the NV^0 state can acquire one electron and turns to NV^-. It's found that NV center's charge state is sensitive to its surrounding charge environments [6, 73]. Along this line, the NV center has been used to probe functionalized groups [74, 75], DNA molecules [76, 77] on diamond surface as well as applied voltage on diamond [73, 42]. Recently, we have shown that metal cations can be detected via charge state transition of NV centers inside nanodiamonds [78]. Details will be discussed in Chapter 5.

With the above-mentioned sensing protocols, the NV system has been utilized in many different fields. Examples include biological sensing [79, 80, 81], such as probing spin-labelled [82] or unlabelled DNA [4], tumor tissue activities [83], molecular transitions [84], embryogenesis processes [85]. Diamond surface chemistry and engineering would provide more opportunities in this direction [86, 87].

Moreover, thanks to its large temperature operation range (from cryogenic temperatures to above room temperature), frequency detection range (spanning from DC to GHz) and excellent spatial resolution (sensor–sample distances as small as a few nanometres), NV center provides access to both static and dynamic magnetic and electronic phenomena with nanoscale resolution, thus being a great candidate of probing condensed matter physics [88]. For example, NV center can be used to probe correlated-electron physics of magnets and superconductors and explore the current distributions in low-dimensional materials [89, 90, 91].

Quantum network

Thanks to multiple long-lived qubits provided by electronic and nuclear spins and an optical interface for entanglement, NV center in diamond is a promising candidate to act as node of quantum network [92]. A quantum network is poised to enable large-scale secure communication as well as distributed quantum computing and simulation by means of shared entangled states over the nodes of the network. In recent years, diamond-based quantum nodes consisting of NV centers and nuclear spins have witnessed rapid development [93, 94], thanks to various properties of NV systems. For example, NV centers feature spin-state-selective optical transitions that can entangle the color center's spin-state with a photonic state. This makes NV center suitable to be a communication qubit, while nuclear spins such as ^{13}C have a long coherence time even at ambient temperature and can serve as memory qubits. To date, heralded [95] and deterministic [96] entanglement generation between remote solid-state qubits as well as multi-node quantum network based on NV system [97] have been achieved.

While the NV center has been a workhorse of quantum network demonstrations, it suffers from low fraction of optical emission at ZPL (around 3 %) and the significant attenuation of its ZPL photon in optical fiber prevents the network extended to long distances. Along this line, a telecom-wavelength photon (tele-photon) interface would be essential to reduce the photon loss for long-distance entanglement generations. This can be solved by performing parametric down conversion to convert the photons into tele-photons [98], or with another solid-state interface [99].

Apart from the challenges in the physical qubit layer, interesting questions arise in the network layer where robust and well-functioning network design is needed to enable the broader capabilities of quantum networks [100, 101, 102, 103]. In particular, the network design determines routing principles that enable effective network responses to simultaneous requests of entanglement generation with given limited quantum capacity, i.e., a small number of available quantum memories possessed by each node. To tackle with this entanglement routing problem, we proposed an effective routing protocol that embeds quantum operations within the algorithmic workflow [104]. These routing designs may also inspire future work that combine quantum information science with classical network theory.

1.3   Thesis overview

In this thesis, we will present our efforts in exploring promising quantum applications based on the NV system, including quantum geometry measurement and quantum sensing. The thesis can thus be divided into two parts: the first half (Chapter 2 and Chapter 3) focuses on measurement of quantum geometry using a single NV center system and discusses its applications in studying exotic gauge field and quantum metrology; in the second part (Chapter 4 and Chapter 5), we will show how NV sensors can be used as quantum sensors to detect various chemical or biological signals.

More specifically, the thesis is organized as follows:

- **Chapter 2** introduces the characterization of geometric properties of quantum states using quantum geometric tensor (QGT). We will first discuss its physical intuitions and to probe the tensor, we focus on the weak periodic modulation method. We then extract the QGT of a three-level system by engineering the triplet ground state of a single NV center.

- **Chapter 3** discusses several applications of QGT introduced in the previous chapter. Firstly, we show how the measurement of QGT can be used to reveal the existence of tensor monopole in a four-dimensional parameter space and the relation between this monopole and tensor gauge field. Secondly, the topological properties of the system is discussed when the chiral symmetry is broken. Finally, we will draw the connection between QGT and quantum multi-parameter estimation problems.

- **Chapter 4** gives an introduction of NV relaxometry and provide a simple model to quantify NV center's longitudinal relaxation (T_1) process. We then show how we can detect magnetic molecules by measuring the T_1 relaxation time of NV centers inside nanodiamonds.

- **Chapter 5** focuses on using NV centers inside nanodiamond to detect biological and chemical processes. The T_1 relaxometry enables us to design various sensors, including rotational Brownian motion sensors and SARS-CoV-2 virus RNA sensors. Then we will show that the charge state of NV centers in nanodiamonds is sensitive to surface charges, which enables us to design alkali cation sensors with the help of chemical engineering.

- **Chapter 6** provides a brief summary and outlook.

Chapter 2

Quantum geometry: theory and measurement

Geometry plays a significant role in modern physics including quantum physics, general relativity and gauge theories. The belief that our world is ultimately geometrical is commonly held when one approaches physics problems. On one hand, the electroweak and strong interactions are unified by Yang-Mills theory [105], where gauge connections are introduced by comparing nearby frames of the internal spaces such as SU(2), and this connection is reduced to the well-known vector potential A_μ in electromagnetic theory; on the other hand, gravity emerges as the local space-time symmetry and is characterized by the Christoffel connection [106]. In quantum systems, gauge fields can emerge in a similar manner as those in fundamental forces between elementary particles [107, 108, 109]. A well-celebrated example is that the motion of electrons or other particles with spins can be modified by an effective gauge field now known as Berry curvature. With fiber bundle theory, quantum states that differ only by a local phase factor can be identified as rays on a complex projective space, which are endowed with the unitary-invariant metric known as Fubini-Study metric as we will show later. The above ideas are then described by a broader formalism of geometric quantum mechanics [110].

In this chapter, we will introduce the quantum geometric tensor, an object which describes the emergent geometric structure in geometric quantum mechanics [110] via the Fubini-Study metric and Berry curvature. We will then demonstrate how we can measure the tensor in a qutrit system synthesized by a single NV center in diamond and present experimental results.

2.1 Quantum geometric tensor

To quantify the geometric information of a quantum state, we rely on the quantum geometric tensor (QGT) [111, 112, 113]. The notion of QGT first appeared in 1980 in Ref. [111] and it's generally defined on any manifold of states smoothly varying with parameter $\boldsymbol{q}$. In this thesis, we will consider non-degenerate ground state manifold of Hamiltonian $\mathcal{H}$ of interest. QGT $\chi_{\mu\nu}$ naturally appears when one defines the distance between nearby states $|n(\boldsymbol{q})\rangle$ and $|n(\boldsymbol{q}+d\boldsymbol{q})\rangle$:

$$ds^2 \equiv 1 - |\langle n(\boldsymbol{q})|\, n(\boldsymbol{q}+d\boldsymbol{q})\rangle|^2 = dq_\mu \chi_{\mu\nu} dq_\nu + O(|d\boldsymbol{q}|^3), \tag{2.1}$$

where we apply the Taylor expansion about $d\boldsymbol{q} = 0$ and assume a generic system parametrized by the generalized position $\boldsymbol{q}$. Here $\chi_{\mu\nu}$ contains information on the geometry of the manifold for the state $|n\rangle$ (labelled as a superscript in equations below) and is found to be:

$$\chi_{\mu\nu}^{(n)} = \langle \partial_\mu n(\boldsymbol{q})|\,(1 - |n(\boldsymbol{q})\rangle \langle n(\boldsymbol{q})|)\,|\partial_\nu n(\boldsymbol{q})\rangle = g_{\mu\nu} + i\mathcal{F}_{\mu\nu}/2. \tag{2.2}$$

Since the quantum states are represented by complex vectors, QGT is a complex tensor: the symmetric part is the Fubini-Study metric tensor $g_{\mu\nu}$ that serves as Riemannian metric and it measures the "quantum distance" between the states; the anti-symmetric part is the conventional 2-form Berry curvature $\mathcal{F}_{\mu\nu}$ and is identified with the emergent gauge field in the projected Hilbert space. An intuitive way to understand Eq. 2.1 is that the inner product between two nearby quantum states contains two part information, the overlap characterized by the metric tensor $g_{\mu\nu}$ and the relative phase characterized by the curvature $\mathcal{F}_{\mu\nu}$.

If we consider $|n(\boldsymbol{q})\rangle$ to be the eigenstate of a Hamiltonian $\mathcal{H}$ with ϵ_k being the eigen-energy of eigenstate $|k(\boldsymbol{q})\rangle$, we can relate $\chi_{\mu\nu}^{(n)}$ to variations of the Hamiltonian. Using time-independent perturbation theory to first order we obtain

$$|\partial_\mu n(\boldsymbol{q})\rangle = \sum_{k \neq n} \frac{\langle k(\boldsymbol{q})| \, \partial_\mu \mathcal{H} \, |n(\boldsymbol{q})\rangle}{\epsilon_n - \epsilon_k} |k(\boldsymbol{q})\rangle \,, \tag{2.3}$$

Then we can plug it in Eq. 2.2 and simplify to the following form

$$
\begin{aligned}
\chi_{\mu\nu}^{(n)} &= \sum_{k \neq n} \sum_{m \neq n} \frac{\langle k| \, \partial_\mu \mathcal{H} \, |n\rangle^\dagger}{\epsilon_n - \epsilon_k} \langle k| \left(1 - |n\rangle \langle n|\right) \frac{\langle m| \, \partial_\nu \mathcal{H} \, |n\rangle}{\epsilon_n - \epsilon_m} |m\rangle \\
&= \sum_{m \neq n} \frac{\langle m| \, \partial_\mu \mathcal{H} \, |n\rangle^\dagger}{\epsilon_n - \epsilon_m} \frac{\langle m| \, \partial_\nu \mathcal{H} \, |n\rangle}{\epsilon_n - \epsilon_m} \\
&= \sum_{m \neq n} \frac{\langle n| \, \partial_\mu \mathcal{H} \, |m\rangle \langle m| \, \partial_\nu \mathcal{H} \, |n\rangle}{(\epsilon_n - \epsilon_m)^2}.
\end{aligned}
\tag{2.4}
$$

The QGT indeed captures the geometric information of the state $|n(\boldsymbol{q})\rangle$ and would therefore be very useful in extracting interesting topological quantities, as we will show later. For example, the Berry curvature is well-known in condensed matter physics and has been used to characterize various topological materials such as Chern insulators. More over, from the definition, as stated above, it's easy to see that the metric tensor characterizes the distance or overlap between two nearby states upon a small perturbation. That is, a larger metric tensor would indicate a larger distinguishability between two states differing by an infinitesimal change of parameters. Therefore, it would play an important role in quantum metrology. The Berry curvature term, instead, is linked to the precision when one considers multiparameter estimation scenarios. We will discuss these relations in detail in the next chapter.

2.2   Measurement of quantum geometric tensor

In this section, we move to the experimental measurement of QGT. The experimental results shown below can also be found in the published journal paper Ref. [114]. We consider a system with Hamiltonian $\mathcal{H}(\boldsymbol{q})$ and the goal is to extract the QGT of its ground state $|n_-\rangle$ with energy ϵ_- (we use the subscript $-$ here to make the notation coherent through the chapters).

2.2.1   Weak periodic modulation of parameters

While QGT can be probed via quantum state tomography given the definition in
Eq. 2.2, it takes exponential long time as the number of parameters increases. Here
we instead introduce a method to measure the QGT using weak modulations of
the parameters $\mu, \nu \in \boldsymbol{q}$ [115, 116]. That is, we set $\mu_t = \mu_0 + m_\mu \sin(\omega t)$, $\nu_t =$
$\nu_0 + m_\nu \sin(\omega t)$ (linear) or $\mu_t = \mu_0 + m_\mu \cos(\omega t)$, $\nu_t = \nu_0 + m_\nu \sin(\omega t)$ (elliptical), with
$m_\mu, m_\nu \ll 1$ being the modulation strength and ω being the modulation frequency.

Quantum metric tensor measurement

Specifically, to extract the real part of the QGT, we first apply the following linear
modulations on the parameters

$$
\begin{aligned}
\mu_t &= \mu_0 + m_\mu \sin \omega t, \\
\nu_t &= \nu_0 + m_\nu \sin \omega t.
\end{aligned}
\tag{2.5}
$$

If we modulate the parameters weakly, $m_\mu, m_\nu \ll 1$, according to perturbation theory

$$
\mathcal{H} \approx \mathcal{H}(\boldsymbol{q}_0) + m_\mu \partial_\mu \mathcal{H} \sin \omega t + m_\nu \partial_\nu \mathcal{H} \sin \omega t.
\tag{2.6}
$$

When the modulation frequency ω is resonant with the energy difference between
the ground state $|n_-\rangle$ and other eigenstates, the parametrically modulated Hamilto-
nian will drive coherent Rabi oscillations between these states. For example, let's con-
sider the case where $\omega = \epsilon_+ - \epsilon_-$ is resonant with the transition between $|n_-\rangle \leftrightarrow |n_+\rangle$.
Here $|n_+\rangle$ is one of the excited state of the system with eigenenergy ϵ_+. Then, we
will have the Rabi frequency

$$
\begin{aligned}
\Omega_{-\leftrightarrow+} &= |\langle n_- | \mathcal{H}(\boldsymbol{q}_0) + m_\mu \partial_\mu \mathcal{H} + m_\nu \partial_\nu \mathcal{H} |n_+\rangle| \\
&= |\langle n_- | m_\mu \partial_\mu \mathcal{H} + m_\nu \partial_\nu \mathcal{H} |n_+\rangle|
\end{aligned}
\tag{2.7}
$$

To measure the diagonal component $g_{\mu\mu}$ of the metric tensor, we set $m_\nu = 0$:

$$\Omega^\mu_{-,+} \equiv m_\mu \Gamma^\mu_{-,+} = m_\mu |\langle n_- | \partial_\mu \mathcal{H} | n_+ \rangle |. \tag{2.8}$$

It's clear to see that this Rabi frequency is the same as the expression in Eq. 2.4. Following the same procedure, by setting the driving frequency ω to be resonant with the transition energy between ground state $|n_-\rangle$ and other eigenstates, we can obtain the contributions from all the excited states.

Now with Eq. 2.4, we can get the diagonal components of the metric tensor from experimentally measurable quantities:

$$g_{\mu\mu} = \sum_{m \neq -} \frac{(\Gamma^\mu_{-,m})^2}{(\epsilon_m - \epsilon_-)^2}. \tag{2.9}$$

To measure the off-diagonal components $g_{\mu\nu}$, we modulate both parameters such that $m_\mu = \pm m_\nu$. Then the coherent Rabi oscillation is:

$$\Omega^{\mu\pm\nu}_{-,+} = m_\mu |\langle n_- | \partial_\mu \mathcal{H} \pm \partial_\nu \mathcal{H} | n_+ \rangle |. \tag{2.10}$$

Setting $\Gamma^{\mu\pm\nu}_{-,+} = \Omega^{\mu\pm\nu}_{-,+} / m_\mu$, we have

$$(\Gamma^{\mu\nu}_{-,+})^2 - (\Gamma^{\mu\bar\nu}_{-,+})^2 = 4|\langle n_- | \partial_\mu \mathcal{H} | n_+ \rangle \langle n_+ | \partial_\nu \mathcal{H} | n_- \rangle | \tag{2.11}$$

Again, with contributions from all other excited states, we thus obtain an expression for the off-diagonal components

$$g_{\mu\nu} = \sum_{m \neq -} \frac{(\Gamma^{\mu\nu}_{-,m})^2 - (\Gamma^{\mu\bar\nu}_{-,m})^2}{4(\epsilon_m - \epsilon_-)^2}. \tag{2.12}$$

Berry curvature measurement

We next present the method measuring the 2-form Berry curvature in QGT. Instead of having a linear modulation on the parameters as in Eq. 2.5, we now use the following elliptical modulation of the system Hamiltonian [115, 116]:

$$\mu_t = \mu_0 + m_\mu \cos \omega t,$$

$$\nu_t = \nu_0 + m_\nu \sin \omega t. \tag{2.13}$$

Again, when the modulation amplitude is small, we have

$$\mathcal{H} \approx \mathcal{H}(\boldsymbol{q}_0) + m_\mu \partial_\mu \mathcal{H} \cos \omega t + m_\nu \partial_\nu \mathcal{H} \sin \omega t, \tag{2.14}$$

Similar as the metric tensor case, we could then measure the Rabi freqeuncy $\langle b_- | \partial_\mu H \pm i \partial_\nu H | m \rangle$, and obtain the imaginary part of the QGT according to Eq. 2.4:

$$\mathcal{F}_{\mu\nu} = \sum_{m \neq -} \frac{[(\Gamma^{\mu\nu}_{-,m})^2 - (\Gamma^{\mu\bar{\nu}}_{-,m})^2]}{2(\epsilon_- - \epsilon_m)^2} \ . \tag{2.15}$$

2.3 Experimental measurement of QGT in a qutrit system

With the method introduced above, we will next present our experimental results in measuring the QGT of a three-level system using a single NV center in diamond. We will consider a specific four-dimensional Hamiltonian $\mathcal{H}(\boldsymbol{q})$ and in Chapter 3 we will show its connection with topological quantities.

Setup and sample

We start by presenting the basic information of sample and setup used in this work. We used a home-built confocal microscope to initialize and measure a single NV center in an electronic grade diamond sample (Element 6, natural abundance of ^{13}C). A magnetic field of 490 G is applied using a permanent magnet, and the native ^{14}N nuclear spin is thus polarized and remains in $|m_I = +1\rangle$ state during the experiments due to the excited state level anti-crossing effect. We show an image of the setup in Fig. 2-1 below.

Figure 2-1: Picture of single NV measurement setup. We use a coplanar waveguide (CPW) to delivery microwave and the objective is below the diamond sample.

The NV of interest has a long coherence time with $T_1 = 3.2$ ms and $T_{2,echo} > 700$ μs. We remark that while the weak periodic modulation method might also be adopted in other quantum platforms such as superconducing circuits, the long coherence time of the NV center allows us to extract the QGT from measuring Rabi oscillations.

In order to have full control on the NV triplet ground state, generation of dual-frequency microwave pulses that are on resonance with $|m_s = 0\rangle \leftrightarrow |m_s = \pm 1\rangle$ is needed. To achieve precise control over the amplitude and phase of both microwave frequencies, we choose to use frequency modulation with two separate IQ mixers, shown schematically in Fig. 2-2. The in-phase (I) and quadrature (Q) RF signals are generated from AWG (Tektronix 5014B) using 3 separate channels. One of the outputs is further split into 0 and 90° (Mini-Circuits ZMSCQ-2-90). A two-channel microwave generator (Windfreak SynthHD) generates the Local Oscillators (LO). The IQs and LOs are combined in two IQ mixers (Texas Instrument, TRF370317; Marki Microwave, IQ-0318) that create up-converted single-sideband microwave signals. These output signals are then combined and controlled by a microwave switch (Analog Devices, ADRF5020) before amplified. For brevity, we left out pre-amplifiers in the schematics. With optical polarization and readout, now we have full control

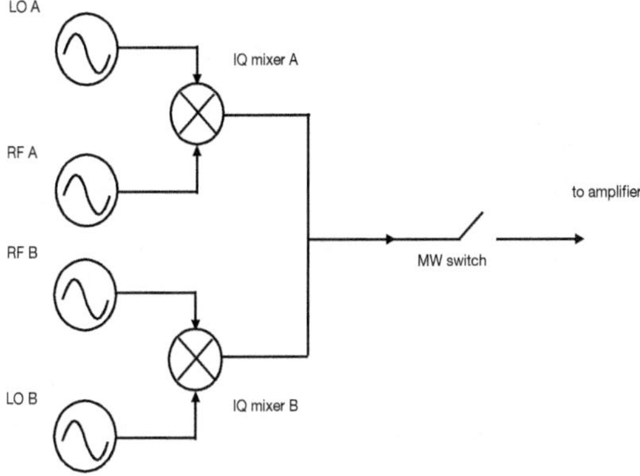

Figure 2-2: Setup schematics. Two separate IQ mixers is used to generate dual-frequency pulse.

on the three-level system.

Hamiltonian generation

We now derive the effective Hamiltonian in the rotating frame of the dual-frequency microwave pulses following Ref. [117]. The intrinsic Hamiltonian of NV is $DS_z^2+\gamma_e BS_z$ (see Fig. 2-3), where $\gamma_e = 2.8$ MHz/G is the gyromagnetic ratio, $D = 2.87$ GHz is the zero-field energy splitting of the NV ground state, $B = 490$ G here is the external field along N-V axis and S_z is the spin-1 z operator. Under the dual-frequency microwave control, the total Hamiltonian is given by:

$$H_{NV} = DS_z^2 + \gamma_e BS_z$$
$$+ 2\sqrt{2}[\gamma_e B_1 \cos(\omega_1 t + \phi_1)S_x + \gamma_e B_2 \cos(\omega_2 t + \phi_2)]S_x \tag{2.16}$$

where S_x, S_z are the spin-1 operators, the first line is the NV spin Hamiltonian and the second line represents the dual-frequency microwave pulse at frequencies ω_1, ω_2. In the bare NV frame with basis $|m_s = +1, 0, -1\rangle$, the above Hamiltonian can be written as:

$$H_{NV} = \begin{pmatrix} D + \gamma_e B & 2\left(\begin{array}{c}\gamma_e B_1 \cos(\omega_1 t + \phi_1) \\ + \gamma_e B_2 \cos(\omega_2 t + \phi_2)\end{array}\right) & 0 \\[1em] 2\left(\begin{array}{c}\gamma_e B_1 \cos(\omega_1 t + \phi_1) \\ + \gamma_e B_2 \cos(\omega_2 t + \phi_2)\end{array}\right) & 0 & 2\left(\begin{array}{c}\gamma_e B_1 \cos(\omega_1 t + \phi_1) \\ + \gamma_e B_2 \cos(\omega_2 t + \phi_2)\end{array}\right) \\[1em] 0 & 2\left(\begin{array}{c}\gamma_e B_1 \cos(\omega_1 t + \phi_1) \\ + \gamma_e B_2 \cos(\omega_2 t + \phi_2)\end{array}\right) & D - \gamma_e B \end{pmatrix}.$$

$$(2.17)$$

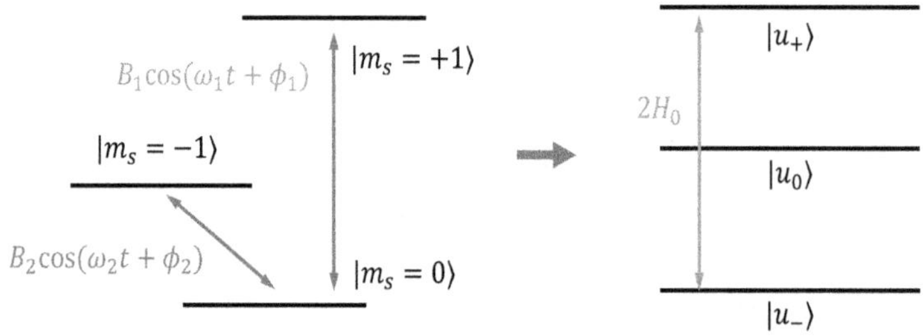

Figure 2-3: The ground state spin degrees of freedom of a single NV center is shown on the left. The spin sub-levels are coupled via a dual-frequency microwave pulse. When both microwave pulses are on resonance, we can engineer equally-spacing levels corresponding to the eigenstates of a four-dimensional Hamiltonian, as shown on the right.

We now enter the rotating frame defined by the unitary transformation:

$$V = \begin{pmatrix} e^{-i\omega_1 t} & 0 & 0 \\ 0 & 1 & 0 \\ 0 & 0 & e^{-i\omega_2 t} \end{pmatrix}.$$

The Hamiltonian Eq. 2.17 can be rewritten as:

$$\mathcal{H} = \begin{pmatrix} D + \gamma_e B - \omega_1 & B_1 e^{-i\phi_1} & 0 \\ B_1 e^{i\phi_1} & 0 & B_2 e^{i\phi_2} \\ 0 & B_2 e^{-i\phi_2} & D - \gamma_e B - \omega_2 \end{pmatrix} = \begin{pmatrix} B_z/\sqrt{2} & H_0 \cos\alpha\, e^{-i\beta} & 0 \\ H_0 \cos\alpha\, e^{i\beta} & 0 & H_0 \sin\alpha\, e^{i\phi} \\ 0 & H_0 \sin\alpha\, e^{-i\phi} & -B_z/\sqrt{2} \end{pmatrix}$$

$$(2.18)$$

where $B_z/\sqrt{2} = D \pm \gamma_e B - \omega_{1(2)}$ corresponds to detunings in microwave frequency, $B_1 = H_0 \cos \alpha$, $B_2 = H_0 \sin \alpha$ and $\phi_1 = \beta, \phi_2 = \phi$ are the amplitudes and phases of the microwave pulses, respectively. We call this Hamiltonian the double quantum (DQ) Hamiltonian in the following, for its ability to drive the $|m_s = -1\rangle \leftrightarrow |m_s = +1\rangle$ transition, where the quantum number changes by 2.

When both microwave frequencies are on-resonance $\omega_{1(2)} = D \pm \gamma_e B$, the eigenstates of Eq. 2.18 are:

$$|u_\pm\rangle = \frac{1}{\sqrt{2}} \begin{pmatrix} \frac{B_1 e^{-i\phi_1}}{\sqrt{B_1^2 + B_2^2}} \\ \pm 1 \\ \frac{B_2 e^{-i\phi_2}}{\sqrt{B_1^2 + B_2^2}} \end{pmatrix}, |u_0\rangle = \begin{pmatrix} \frac{B_2 e^{-i\phi_1}}{\sqrt{B_1^2 + B_2^2}} \\ 0 \\ \frac{-B_1 e^{-i\phi_2}}{\sqrt{B_1^2 + B_2^2}} \end{pmatrix}, \tag{2.19}$$

and the corresponding eigen-energies are $\epsilon_\pm = \pm \gamma_e \sqrt{B_1^2 + B_2^2}$ and $\epsilon_0 = 0$. The energy levels are shown in Fig. 2-3. In this case, we have a Weyl-type Hamiltonian as the spectrum is similar as the momentum space spectrum of Weyl semimetals.

To characterize our engineered Hamiltonian in Eq. 2.18 under dual-frequency microwave driving, we prepare NV in the $|m_s = 0\rangle$ state and let it evolve under the DQ Hamiltonian (Eq. 2.18). When both microwave frequencies are on-resonance, we expect the following time-dependent state evolution

$$\begin{pmatrix} c_+(t) \\ c_0(t) \\ c_-(t) \end{pmatrix} = \begin{pmatrix} \frac{-iB_1 e^{-i\phi_1} \sin \omega_e t}{\sqrt{B_1^2 + B_2^2}} \\ \cos \omega_e t \\ \frac{-iB_2 e^{-i\phi_2} \sin \omega_e t}{\sqrt{B_1^2 + B_2^2}} \end{pmatrix}, \tag{2.20}$$

with the effective Rabi frequency $\omega_e = \gamma_e \sqrt{B_1^2 + B_2^2}$. By measuring the amplitude and frequency of the Rabi oscillation, we can extract both B_1 and B_2. Here, we work in the linear regime of the microwave amplifier, where $\omega_e = 2$ MHz. We show the evolution of the three states $|m_s = 0, \pm 1\rangle$ when the parameter $\alpha = 0, \pi/4$.

Indeed, when $B_z = 0$, the Hamiltonian Eq. 2.18 has four independent parameters and holds a topological charge at the origin. We will leave the discussion of the

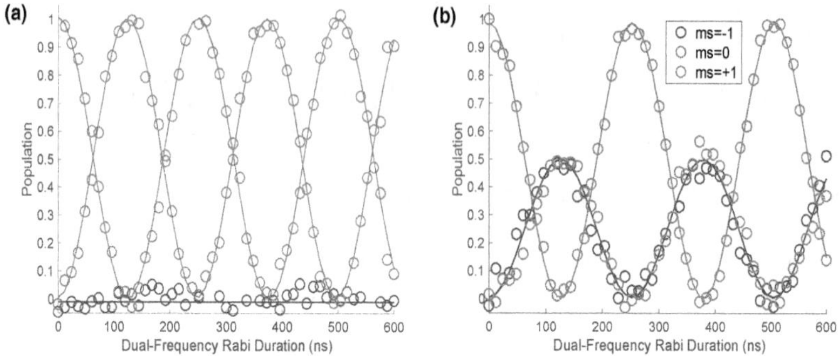

Figure 2-4: Evolution of the state $|m_s = 0, \pm 1\rangle$ under the DQ Hamiltonian in Eq. 2.18. The experimental conditions are **(a)** $\alpha = 0$ and **(b)** $\alpha = \pi/4$. The expected oscillation amplitude for the $|m_s = +1\rangle$ $(|m_s = -1\rangle)$ state is $\cos^2 \alpha$ $(\sin^2 \alpha)$, in good agreement with our experiments

topological properties to the next chapter and now focus on the measurement of QGT of its ground state $|u_-\rangle$.

Ground state preparation and readout

In order to probe the QGT, we need to first prepare the ground state. At the beginning of every parametric modulation experiment, we polarize the NV into $|m_s = 0\rangle$, then apply two microwave pulses to prepare NV into the ground eigenstate $|u_-\rangle$ before subjecting the system to the engineered Weyl-type Hamiltonian. Again, the ground state can be written as

$$|u_-\rangle = \frac{1}{\sqrt{2}} \begin{pmatrix} -\cos\alpha e^{-i\beta} \\ 1 \\ -\sin\alpha e^{-i\phi} \end{pmatrix}. \tag{2.21}$$

We first apply a microwave pulse that drives the -1 transition $|m_s = 0\rangle \leftrightarrow |m_s = -1\rangle$, immediately followed by another pulse for the $+1$ transition $|m_s = 0\rangle \leftrightarrow |m_s = +1\rangle$. We set the Rabi frequency of both pulses to be ω_{init}. Then the durations $t_\pm$ and phases $\delta_\pm$ of the two initialization pulses for the ± 1 transitions are

$$
\begin{aligned}
t_- &= \sin^{-1}(\sin\alpha/\sqrt{2})/\omega_{init}, \\
\delta_- &= \phi + \pi/2, \\
t_+ &= \sin^{-1}(\cos\alpha/\sqrt{2-\sin^2\alpha})/\omega_{init}, \\
\delta_+ &= \beta + \pi/2.
\end{aligned}
\tag{2.22}
$$

Similarly, if the goal is to prepare the first excited state $|u_0\rangle = (-\sin\alpha e^{-i\beta}, 0, \cos\alpha e^{-i\phi})^T$, the parameters are

$$
\begin{aligned}
t_- &= \pi/\omega_{init}, \\
\delta_- &= \phi + \pi, \\
t_+ &= \pi/2\omega_{init}, \\
\delta_+ &= \beta.
\end{aligned}
\tag{2.23}
$$

After successfully preparing the state into $|u_-\rangle$, the system then undergoes evolution under the engineered Hamiltonian Eq. 2.18 and Rabi oscillations occur between ground state and one of the excited states. Then, to probe the population of the ground state, we apply the inverse mapping to rotate $|u_-\rangle$ back to $|m_s = 0\rangle$ for fluorescent readout. While the case when the modulation frequency $\omega_r = 2H_0$ is simple, it's more subtle when the modulation frequency is resonant with $|u_0\rangle \leftrightarrow |u_\pm\rangle$ transitions. Due to chiral symmetry of the Weyl-type Hamiltonian, they have the same driving strength $\Gamma = |\Gamma_{-,0}| = |\Gamma_{+,0}|$. This leads to an effective DQ Hamiltonian in the eigenbasis of our engineered Weyl-type Hamiltonian. In the DQ rotating frame and after taking the rotating wave approximation similar to what we did in Eq. 2.17, we obtain the following Hamiltonian in the eigenbasis of Eq. 2.18 (whose eigenvectors are $|u_{0,\pm}\rangle$):

$$
\hat{H} = \begin{pmatrix} 0 & \Gamma e^{-i\phi_1} & 0 \\ \Gamma e^{i\phi_1} & 0 & \Gamma e^{i\phi_2} \\ 0 & \Gamma e^{-i\phi_2} & 0 \end{pmatrix},
\tag{2.24}
$$

where $\phi_{1(2)}$ are the phases associated with $\Gamma_{0,\pm}$. Then, starting from the ground state $|u_-\rangle$, the system under Eq. 2.24 evolves as:

$$\begin{pmatrix} c_+(t) \\ c_0(t) \\ c_-(t) \end{pmatrix} = \begin{pmatrix} \frac{(-1+\cos(\sqrt{2}\Gamma t))e^{i(\phi_1-\phi_2)}}{2} \\ \frac{1}{\sqrt{2}}\sin(\sqrt{2}\Gamma t)e^{i\phi_1} \\ \frac{1+\cos(\sqrt{2}\Gamma t)}{2} \end{pmatrix}. \tag{2.25}$$

For ease of fitting, we map $|u_0\rangle$ back to $|m_s=0\rangle$ and read out the population optically. The matrix element Γ of interest is $1/\sqrt{2}m_\mu$ of the fitted Rabi frequency.

We remark that the mapping pulses described in this section do not perform the unitary transformation between the basis $\{|m_s\rangle\}$ and $\{|u\rangle\}$. We emphasize that this unitary transformation could be achieved by three microwave pulses, and is useful in determining the relevant matrix element when the chiral symmetry of the Hamiltonian is broken by adding a non-zero B_z term in Eq. 2.18, where $|\Gamma_{-,0}| \neq |\Gamma_{+,0}|$, and the modulation frequency is resonant for both SQ transitions. In this case, we prepare the initial state in $|u_0\rangle$, such that it evolves according to Eq. 2.20. By measuring both the amplitude and frequency of the Rabi oscillation in $|u_\pm\rangle$ states, we are able to reconstruct the matrix element of interest.

We remark that to determine the oscillation amplitudes accurately for all three states, we have to perform three sets of experiments. Each experiment consists of the same state preparation, parametric modulation, and the unitary map back to all three $|m_s\rangle$ states, followed by (i) no operation (ii) π pulse between $|m_s=0\rangle \leftrightarrow |m_s=-1\rangle$ and (iii) π pulse between $|m_s=0\rangle \leftrightarrow |m_s=+1\rangle$, and then optically read out. The fluorescence signals recorded in each experiments are labelled as S_i, the reference fluorescent level for each $|m_s\rangle$ states as measured in separate experiments are r_{m_s}, and the final population of each $|m_s\rangle$ states are n_{m_s}. From the three sets of experiment, we have

$$r_+n_+ + r_0n_0 + r_-n_- = S_1,$$
$$r_+n_+ + r_-n_0 + r_0n_- = S_2, \tag{2.26}$$
$$r_0n_+ + r_+n_0 + r_-n_- = S_3.$$

It is therefore straightforward to extract the populations accurately

$$
\begin{pmatrix} n_+ \\ n_0 \\ n_- \end{pmatrix} = \begin{pmatrix} r_+ & r_0 & r_- \\ r_+ & r_- & r_0 \\ r_0 & r_+ & r_- \end{pmatrix}^{-1} \begin{pmatrix} S_1 \\ S_2 \\ S_3 \end{pmatrix} . \tag{2.27}
$$

In addition to the parametric modulations, this readout technique is used in e.g. Fig. 2-4 to reveal the accurate populations of all three states. We remark that this method is only possible when the magnetic field is close to the excited state level anticrossing at around 510 G, where the excited state electron-nuclear spin flip-flops yield distinguishable fluorescent levels for $|m_s = \pm 1\rangle$ [118].

Determination of resonance frequency and measurement of Rabi osciall-tions

We start our measurements by precisely determining the resonant frequency $\omega_r = \epsilon_+ - \epsilon_- = 2(\epsilon_0 - \epsilon_-)$ (when $B_z{=}0$). As shown in Fig. 2-5, we fix the time and sweep the modulation frequency ω to find the resonance condition. We choose a very weak modulation amplitude to reduce power broadening and improve the precision in estimating ω_r.

We then measure the coherent Rabi oscillations under linear parametric modulations at the calibrated $\omega = \omega_r/2$ (ω_r) for transitions between $|u_-\rangle$ and $|u_{0,+}\rangle$ (labelled as SQ and DQ transition hereafter). Examples of SQ and DQ Rabi curves are shown in Fig. 2-6, which includes both single- and two-parameter modulations, associated with measuring the diagonal and off-diagonal components of the metric tensor.

For every combination of modulations $\mu(\mu\nu)$, we measure both the SQ and DQ Rabi frequencies and recover the matrix element $\Gamma^{\mu}_{-,n}$ ($\Gamma^{\mu\nu}_{-,n}$). In Fig. 2-7 and Fig. 2-8, we show measured matrix elements Γ for linear and elliptical modulation respectively, showing good agreement with theoretical predictions.

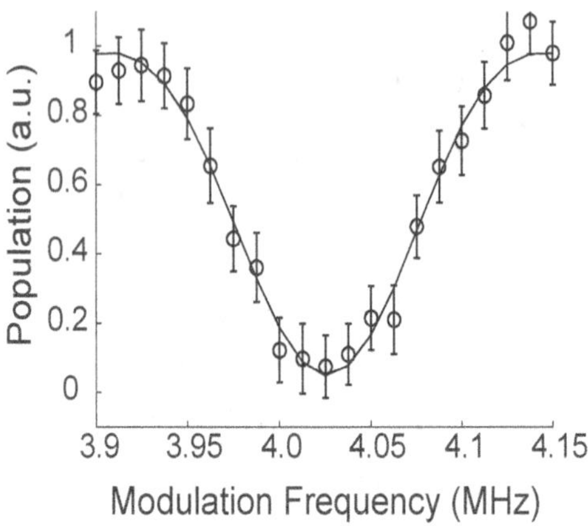

Figure 2-5: To determine the resonance condition for parametric modulation, we fix $\tau = 7.5\mu s$ and the modulation strength on the three parameters in Eq. 2.18 are $(m_\alpha, m_\beta, m_\phi) = (0, 1/30, 1/30)$. We sweep the modulation frequency around 4 MHz to find $\omega_r = 2H_0$.

Extraction of QGT

From the relation between these matrix elements with the metric tensor and the 2-form Berry curvature discussed in Sec. 2.2, we can now reconstruct the whole QGT for the state $|n_-\rangle$, as shown in Fig. 2-9. We remark that $g_{\mu\nu} = g_{\nu\mu}$, and there are in total 6 independent metric tensor components in the 4D parameter space. As the Berry curvature is anti-symmetric, there are only three components in the 4D parameter space and the $\mathcal{F}_{\phi\beta}$ term is always zero.

We note that the Weyl-type Hamiltonian in Eq. 2.18 (at B_z=0) are rotationally symmetric about β, ϕ under our parametrization of $(H_0, \alpha, \beta, \phi)$. As a result, the metric tensor and generalized 3-form Berry curvature is independent of β, ϕ. To further show that this indeed the case in our experiments, we fix $\alpha = \pi/8$ and sweep $\beta, \phi \in [0, 2\pi]$, and measure the corresponding matrix element Γ for the metric tensor components $g_{\alpha\alpha}, g_{\beta\phi}$. The metric tensor results are shown in Fig. 2-10, where we indeed see these measurements remain constant within experimental error for different β, ϕ.

The measured results in Fig. 2-9 have a great agreement with theoretical calcula-

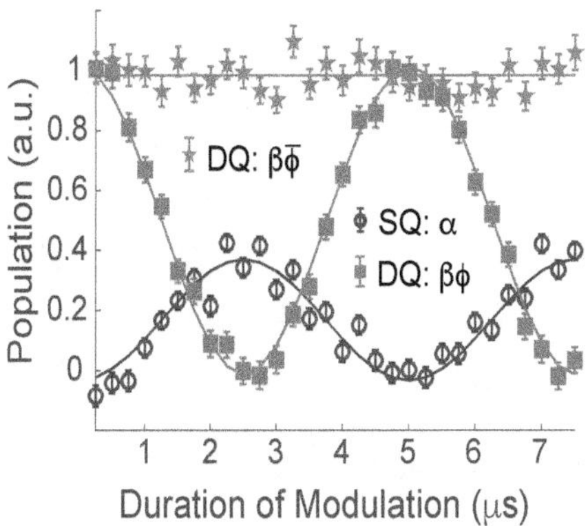

Figure 2-6:   Examples of coherent Rabi oscillations observed under parametric modulations, for the engineered Hamiltonian at ($\alpha_0 = \pi/4, \beta_0 = \phi_0 = 0$). Markers are experimental data and solid curves are fittings. The Rabi frequencies are used to calculate the matrix elements $\Gamma^{\mu(\nu)}_{-,m}$. To extract the diagonal components of the metric tensor, we use a single-parameter modulation, as shown, e.g. by the blue curve, representing the SQ transition ($\omega = \omega_r/2$) for α modulation. Due to chiral symmetry, $|\Gamma^{\mu(\nu)}_{-,0}| = |\Gamma^{\mu(\nu)}_{+,0}|$. We therefore measure the population in the first excited state $|u_0\rangle$, which gives half contrast. The other two curves represent two-parameter modulations resonant with the DQ transition ($\omega = \omega_r$), and possess full contrast.

tions that

$$\chi = g + i\frac{\mathcal{F}}{2} = \begin{pmatrix} \frac{1}{2} & \frac{i\sin(2\alpha)}{4} & -\frac{i\sin(2\alpha)}{4} \\ -\frac{i\sin(2\alpha)}{4} & \frac{\cos^2(\alpha)[2-\cos^2(\alpha)]}{4} & -\frac{\sin^2(2\alpha)}{16} \\ \frac{i\sin(2\alpha)}{4} & -\frac{\sin^2(2\alpha)}{16} & \frac{\sin^2(\alpha)[2-\sin^2(\alpha)]}{4} \end{pmatrix}. \tag{2.28}$$

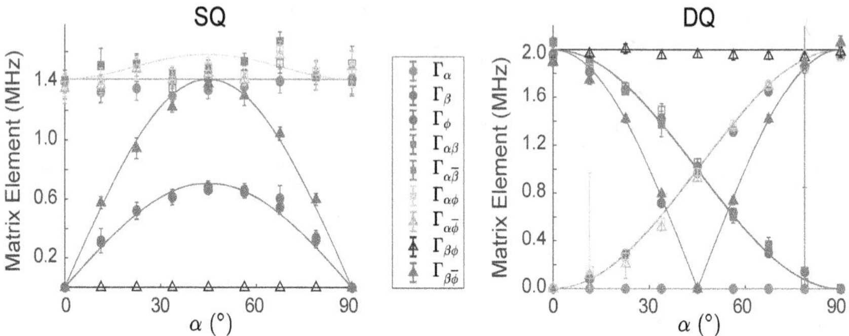

Figure 2-7: Matrix elements $|\Gamma^{\mu(\nu)}_{-,0}|$ measured for SQ (left, $\omega = \omega_r/2$) and $|\Gamma^{\mu(\nu)}_{-,+}|$ for DQ (right, $\omega = \omega_r$) transitions under linear modulation when $B_z = 0$. Note that many matrix elements are expected from theory to coincide and thus their measured values are superimposed at 2 MHz. Markers are experimental data and solid lines are theory.

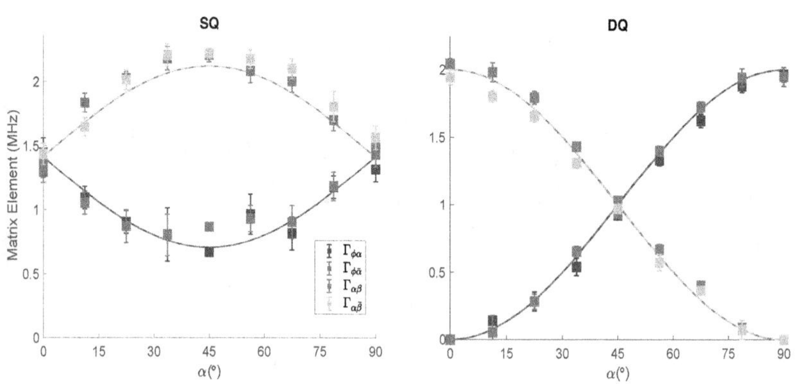

Figure 2-8: Matrix elements $|\Gamma^{\mu(\nu)}_{-,0}|$ measured for SQ (left, $\omega = \omega_r/2$) and $|\Gamma^{\mu(\nu)}_{-,+}|$ for DQ (right, $\omega = \omega_r$) transitions under elliptical modulation when $B_z = 0$. Markers are experimental data and solid lines are theory.

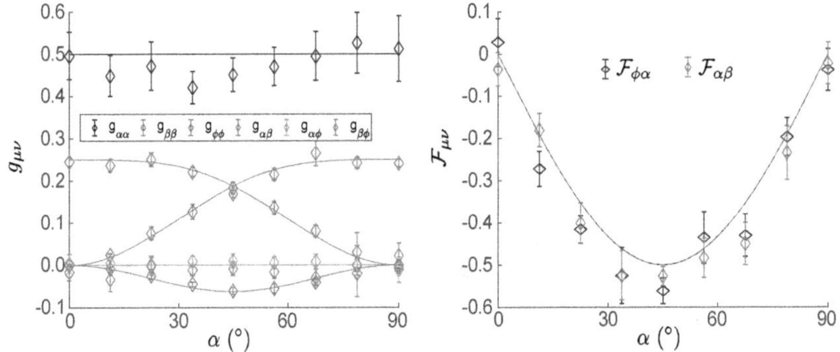

Figure 2-9: Left shows all 6 independent components of the metric tensor as functions of α and right figure is the measurements of non-zero 2-form Berry curvature as function of α.

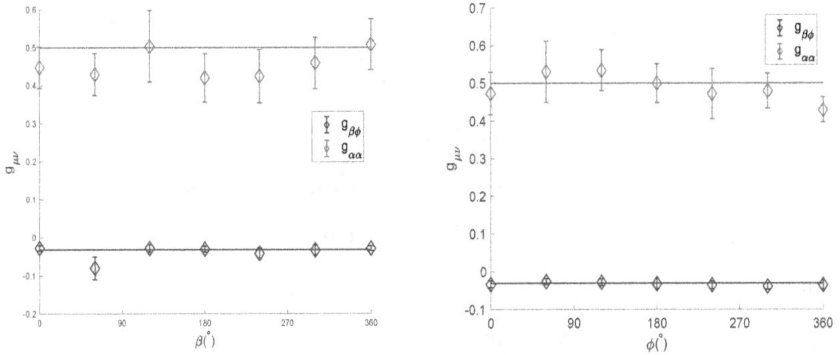

Figure 2-10: To verify that the metric tensor is independent of β, ϕ due to the rotation symmetry of the Hamiltonian, we perform experiments at $\alpha = \pi/8$, sweeping either β or ϕ and measuring $g_{\alpha\alpha}, g_{\beta\phi}$. On the left (right) we show extracted metric tensor components when fixing $\phi = 0$ ($\beta = 0$) and sweeping β (ϕ). Within experimental error they stay constant over β, ϕ.

Chapter 3

Quantum geometry: applications

The measurement of QGT helps us explore geometry-related applications of the four-dimensional model in Eq. 2.18. These applications would be the focus of this chapter.

3.1 Simulation of tensor gauge field

As the first application, we will show how we can connect QGT with simulation of tensor gauge fields that are important in string theory [114]. Our current understanding of fundamental physical phenomena relies on two main pillars, general relativity and quantum field theory. Their mutual incompatibility, however, poses critical limitations to the formulation of a unifying theory of all fundamental interactions. String theory proposes a powerful and elegant formalism to unify gravitational and quantum phenomena, providing a concrete route to quantum gravity [119]. Within this scenario, conventional point-like particles are replaced by extended objects, such as closed and open strings, and conventional vector gauge fields are promoted to tensor (Kalb-Ramond) gauge fields [120, 121]. In direct analogy with the Dirac monopole [122], tensor gauge fields can emanate from point-like defects called tensor monopoles. In four spatial dimensions, the tensor monopole charge is quantized according to the topological Dixmier-Douady ($\mathcal{DD}$) invariant [123, 124, 125], which generalizes the Chern number associated with the Dirac monopole.

Experimental evidence of magnetic monopoles is still lacking in high-energy-physics experiments. However, synthetic monopoles associated with effective gauge fields have recently been detected in quantum systems including ultracold matter and superconducting circuits [126, 127, 128, 129, 116, 130]. Besides, momentum-space monopoles play a central role in topological matter, in particular, characterizing 3D Weyl semimetals. Recently, the notions of tensor monopoles and $\mathcal{DD}$ invariants were shown to arise in 3D chiral topological insulators [131, 132] and in higher-order topological insulators [133].

In the following sections, we will present the experimental observation of the tensor monopole in a minimal synthetic system with our NV system. The results are based on Ref. [114] and we note that the experimental demonstration could also be realized using other systems such as superconducting qubit [134]. We start by giving an introduction of tensor gauge field emanated from a tensor monopole, followed by a comparison between tensor monopole and the celebrated Dirac monopole. In the second part of the section, we move to experimental characterizations and show the measurement of the associated $\mathcal{DD}$ topological invariant.

3.1.1   Tensor monopole and Kalb-Ramond field

Recall that a nodal point in a 3D parameter space is associated with an effective Dirac monopole. In electromagnetism, the Dirac monopole is closely related to the first Chern number. Similar to Gauss's law for the electric charge, the Dirac monopole is revealed by integral of the well-known Berry curvature over any enclosed manifold containing the monopole:

$$C_1 = \frac{1}{2\pi} \int_{S^2} \mathcal{F}_{\mu\nu} dq_\mu \wedge dq_\nu. \tag{3.1}$$

The Berry curvature $\mathcal{F}_{\mu\nu}$ appears in the QGT (Eq. 2.2). In this scenario, the Berry-curvature field emanates radially from the node, and its flux through a 2-sphere enclosing it is quantized, characterized by the Chern number above. This field is indeed derived from a vector gauge field A_μ: $\mathcal{F}_{\mu\nu} = \partial_\mu A_\nu - \partial_\nu A_\mu$.

In a 4D parameter space, the topological charge associated with a nodal point

Quantum geometry	Electromagnetism	Tensor (2-form) gauge field
Berry connection: $A_\mu, B_{\mu\nu}$	vector potential A_μ	tensor potential $B_{\mu\nu}$
Berry curvature: $\mathcal{F}_{\mu\nu}, \mathcal{H}_{\mu\nu\lambda}$	field strength $\mathcal{F}_{\mu\nu}$	field strength $\mathcal{H}_{\mu\nu\lambda}$
Topological invariant	first Chern number C_1	Dixmier-Douady invariant $\mathcal{DD}$

Table 3.1: Ground state quantum geometric properties in electromagnetism in 3D and in the tensor gauge field in 4D.

is provided by a similar invariant, which now involves the flux of a radial 3-form curvature over a 3-sphere that surrounds the node [131]. That is, the 2-form Berry curvature $\mathcal{F}_{\mu\nu}$ is naturally generalized to 3-form curvature $\mathcal{H}_{\mu\nu\lambda}$, which is well known in the context of p-form electromagnetism [121] and is derived from a 2-form gauge field: the Abelian and antisymmetric Kalb-Ramond (KR) field $B_{\mu\nu}$ [120],

$$\mathcal{H}_{\mu\nu\lambda} = \partial_\mu B_{\nu\lambda} + \partial_\nu B_{\lambda\mu} + \partial_\lambda B_{\mu\nu}. \tag{3.2}$$

This KR field plays an important role in string theory, as it naturally couples to extended objects [120, 121].

Now, similarly to the Dirac monopole associated with vector gauge field in 3D space, the tensor KR field $B_{\mu\nu}$ here gives rise to tensor monopoles [123, 124, 125, 131, 132] that are point-like sources of the generalized "magnetic" field $\mathcal{H}_{\mu\nu\lambda}$. To obtain topological charge, one can measure the corresponding flux over a 3-sphere surrounding the monopoles,

$$\mathcal{DD} = \frac{1}{2\pi^2} \int_{S^3} \mathcal{H}_{\mu\nu\lambda} dq^\mu \wedge dq^\nu \wedge dq^\lambda. \tag{3.3}$$

This topological number $\mathcal{DD}$ is known as Dixmier-Douady invariant, which is a generalization of the well-known Chern numbers.

As a comparison, we compare the quantum geometric properties in electromagnetism in 3D and in the tensor gauge field in 4D in Table 3.1.

3.1.2 Model and observables

To engineer such a 4D parameter space, we consider the following Weyl-like Hamiltonian [131],

$$\mathcal{H}_{4D} = \begin{pmatrix} 0 & q_x - iq_y & 0 \\ q_x + iq_y & 0 & q_z + iq_w \\ 0 & q_z - iq_w & 0 \end{pmatrix}. \tag{3.4}$$

We note that this target Hamiltonian is indeed identical to the Hamiltonian than we engineered using the triplet ground state of a single NV center in Eq. 2.18 when both microwave pulses are on resonance, and the ground state of this 4D Hamiltonian is $|u_-\rangle$ with energy ϵ_-. That means that the parameters q=(q_x, q_y, q_z, q_w) here can be expressed in terms of the experimentally controllable parameters $(H_0, \alpha, \beta, \phi)$ through

$$\begin{aligned} q_x + iq_y &= H_0 \cos \alpha e^{i\beta}, \\ q_z + iq_w &= H_0 \sin \alpha e^{i\phi}, \end{aligned} \tag{3.5}$$

where $\alpha \in [0, \pi/2]$ and $\beta, \phi \in [0, 2\pi)$.

The model Eq. 3.4 is the minimal model that holds a tensor monopole at the origin of the parameter space, $\mathbf{q} = 0$. Also we note that this singularity is topologically protected by chiral symmetry $\{U, \mathcal{H}_4 D\} = 0$, where U=diag(1,-1,1). This model further has two mirror symmetries, as we will illustrate later.

According to Eq. 3.3, the field $\mathcal{H}_{\mu\nu\lambda}$ radially emanates from the singularity in 4D space, hence providing an observable and unambiguous signature of tensor monopoles. We now explain how this field can be measured in engineered systems. One may note that the KR field $B_{\mu\nu}$ can be reconstructed from the eigenstates of the Hamiltonian Eq. 3.4. Although state-tomography could be performed to reconstruct these states and the related tensor fields, this approach is resource-intensive. Here we provide two alternative methods to experimentally measure the 3-form curvature $\mathcal{H}_{\mu\nu\lambda}$.

Approach one: metric tensor

The first approach builds on a relation between the 3-form curvature and the quantum metric tensor $g_{\mu\nu}$ that is the real part of the QGT (Eq. 2.2), as illustrated in Ref. [131]:

$$\mathcal{H}_{\mu\nu\lambda} = \epsilon_{\mu\nu\lambda}(4\sqrt{det(g_{\bar{\mu}\bar{\nu}})}), \tag{3.6}$$

where $\bar{\mu}, \bar{\nu} = \{\mu, \nu, \lambda\}$ and $\epsilon_{\mu\nu\lambda}$ is the Levi-Civita symbol. This relation is indeed a generalization of the relation between 2-form Berry curvature and the metric tensor $g_{\mu\nu}$ [131]:

$$\mathcal{F}_{\mu\nu} = 2\epsilon_{\mu\nu}\sqrt{\det g_{\bar{\mu}\bar{\nu}}}, \tag{3.7}$$

where $\epsilon_{\mu\nu}$ is again the Levi-Civita symbol. In short, the relation Eq. 3.6 above allows the measurement of the 3-form curvature through the metric tensor.

Approach one: 2-form Berry curvature

The second approach indeed builds on our experimental parametrization $(H_0, \alpha, \beta, \phi)$. Note that the 2-form tensor connection can be constructed from the state $|u_-\rangle$:

$$B_{\mu\nu} = \Phi\mathcal{F}_{\mu\nu}, \Phi = \frac{-i}{2}\log(u_1 u_2 u_3) \tag{3.8}$$

with $u_{1(2,3)}$ denoting the components of $|u_-\rangle$, $\mathcal{F}_{\mu\nu} = \partial_\mu A_\nu - \partial_\nu A_\mu$ being the 2-form Berry curvature. From Eq. 3.2 and 3.8, it follows that in general, the generalized curvature can be obtained by performing state tomography on the eigenstate, upon external perturbations of the parameters.

With the parameterization given in Eq. 2.18, We can easily prove that, up to a global phase, the ground state of the Hamiltonian has the form

$$|u_-\rangle = [e^{-i\beta}v_1, v_2, e^{-i\phi}v_3]^T \tag{3.9}$$

where v_1, v_2, v_3 are functions of α only. Then, we can calculate the vector gauge potential for this special gauge $\mathcal{A}_\alpha = 0, \quad \mathcal{A}_\beta = v_1^2, \quad \mathcal{A}_\phi = v_3^2$ and the 2-form

Berry curvature $\mathcal{F}_{\mu\nu} = \partial_\mu A_\nu - \partial_\nu A_\mu$

$$\mathcal{F}_{\alpha\beta} = \partial_\alpha(v_1^2),$$
$$\mathcal{F}_{\phi\alpha} = -\partial_\alpha(v_3^2). \tag{3.10}$$

We repeat Eq. 3.2 and 3.8 here for convenience

$$\mathcal{H} = \partial_\alpha B_{\beta\phi} + \partial_\phi B_{\alpha\beta} + \partial_\beta B_{\phi\alpha}, \tag{3.11}$$

where $B_{\mu\nu} = \mathcal{F}_{\mu\nu}\Phi$, and

$$\Phi = -\frac{i}{2}\log\prod_{i=1}^{3} u_i = -\frac{i}{2}\log\left(e^{-i(\phi+\beta)}v_1 v_2 v_3\right).$$

Combining all above, we obtain the simplified form for the curvature

$$\begin{aligned}
\mathcal{H}_{\alpha\beta\phi} &= \partial_\phi(\Phi\mathcal{F}_{\alpha\beta}) + \partial_\beta(\Phi\mathcal{F}_{\phi\alpha}) + \partial_\alpha(\Phi\mathcal{F}_{\beta\phi}) \\
&= -\frac{1}{2}(\mathcal{F}_{\alpha\beta} + \mathcal{F}_{\phi\alpha}) \\
&= -\frac{1}{2}\frac{d}{d\alpha}(v_1^2 - v_3^2),
\end{aligned} \tag{3.12}$$

Now we see from the second line above that by measuring the 2-form Berry curvature (which is related to the imaginary part of QGT), we will be able to extract the field $\mathcal{H}_{\alpha\beta\phi}$.

As a side remark, we note that apart from the construction in Eq. 3.8, the 2-form Berry connection $B_{\mu\nu}$ can be more generally constructed from a mixed set of pseudoreal and complex scalar fields $\psi_{1,2,3}$ satisfying the U(1) gauge transformation and linked to the ground state [114, 132],

$$B_{\mu\nu} = \frac{i}{3}\sum_{j,k,l=1}^{3} \epsilon^{jkl}\psi_j\,\partial_\mu\psi_k\,\partial_\nu\psi_l = \frac{i}{3}\psi_j(\partial_\mu\psi_k\,\partial_\nu\psi_l - \partial_\mu\psi_l\,\partial_\nu\psi_k) \tag{3.13}$$

Choosing for example

$$\psi_1 = -i\log(u_1 + u_3), \qquad \psi_2 = u_1^* - u_3^*, \qquad \psi_3 = u_3 - u_1, \qquad (3.14)$$

where $|u_-\rangle = [u_1, u_2, u_3]^T$, yields a gauge-invariant $\mathcal{H}_{\mu\nu\lambda}$ that would give same observables including $\mathcal{DD}$ as above.

3.1.3 Experimental measurement of $\mathcal{DD}$ invariant

In Chapter 2 we present our measurement results of the whole QGT, including both the metric tensor and 2-form Berry curvature, of the Weyl-like Hamiltonian's ground state $|u_-\rangle$.

Firstly, with Eq. 3.6, we can connect the measured metric tensor to the 3-form curvature that is a generalized "magnetic" field emanated from the tensor monopole in the 4D parameter space. The measured 3-form curvature $\mathcal{H}_{\alpha\beta\phi}$ is shown in Fig. 3-1 (left). The 3-form Berry curvature indeed has an analytical form:

$$\mathcal{H}_{\alpha\beta\phi} = \cos\alpha\sin\alpha. \qquad (3.15)$$

With these data, we obtain the quantization of the generalized "magnetic" flux over the 3-sphere

$$\mathcal{DD}_{expt} = \frac{1}{2\pi^2} \int_0^{\frac{\pi}{2}} d\alpha \int_0^{2\pi} d\beta \int_0^{2\pi} d\phi \, \mathcal{H}_{\alpha\beta\phi} = 0.99(3). \qquad (3.16)$$

which provides an estimation of the $\mathcal{DD}$ invariant in Eq. 3.3 and signals the presence of the tensor monopole at the center of our 4D parameter space.

Alternatively, as discussed above, the tensor monopole can be identified via the 2-form Berry curvatures $\mathcal{F}_{\mu\nu}$ via Eq. 3.12. From the measured values in Fig. 2-9, we can get the reconstructed 3-form curvature in Fig. 3-1 (right). This second approach, which is complementary to the metric-tensor measurement, further confirms the ex-

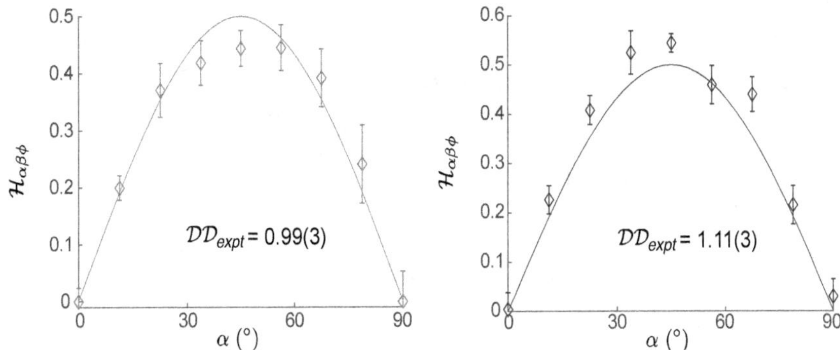

Figure 3-1: Generalized 3-form Berry curvature $\mathcal{H}_{\alpha\beta\phi}$ with respect to α, calculated from the metric tensor (left) and gauge potential (right) using Eq. 3.6 and Eq. 3.12, respectively. The topological invariant $\mathcal{DD}_{expt} = 0.99(3)$ and $1.11(3)$ reveals the existence of a tensor monopole within the hypersphere. Diamonds are experimental data and solid lines are theory. The errorbars are propagated from fitting error of resonant frequencies and Rabi oscillations.

istence of the tensor monopole through the measurement of its quantized charge:

$$\mathcal{DD}_{expt} = \frac{1}{2\pi^2} \int_0^{\frac{\pi}{2}} d\alpha \int_0^{2\pi} d\beta \int_0^{2\pi} d\phi\, \mathcal{H}_{\alpha\beta\phi} = 1.11(3). \tag{3.17}$$

Besides its topological charge, the 4D tensor monopole is also fully characterized by its field distribution [124, 131, 132]

$$\mathcal{H}_{\mu\nu\lambda}(\mathbf{q}) = \epsilon_{\mu\nu\gamma\lambda} q_\gamma / (q_x^2 + q_y^2 + q_z^2 + q_w^2)^2, \tag{3.18}$$

which reflects the fact that the curvature field radially emanates from the topological defect in 4D parameter space. As a consequence, the monopole field has a characteristic inverse-cube dependence on the radial coordinate, that is

$$\mathcal{H} \sim (1/H_0)^3, \tag{3.19}$$

as shown in Fig. 3-2. This signature can also be verified analytically. For convenience, we choose to view the curvature field in cartesian coordinates, (q_x, q_y, q_z, q_w). The radial components of the curvature field satisfy

$$\mathcal{H}^{\perp}_{xyzw}dq_x dq_y dq_z dq_w = \mathcal{H}_{\alpha\beta\phi}\,\det[J]\,dq_x dq_y dq_z dq_w = \frac{1}{H_0^3}dq_x dq_y dq_z dq_w, \qquad (3.20)$$

where J is the Jacobian and we have

$$\det J = \frac{1}{H_0^3}\frac{2}{\sin 2\alpha}. \qquad (3.21)$$

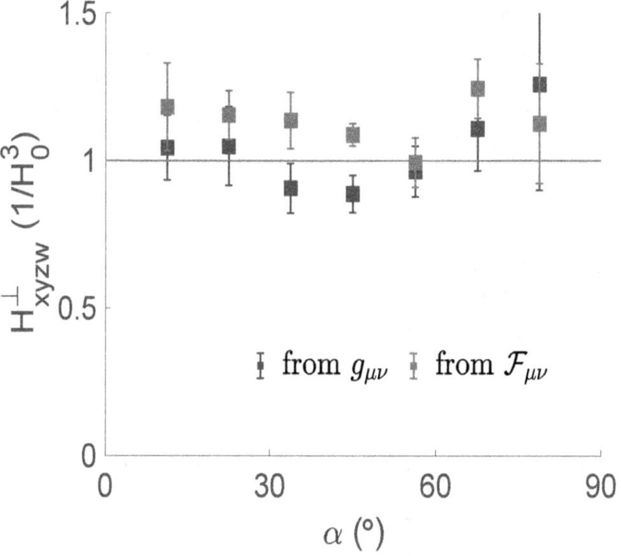

Figure 3-2:   Radial field component $H^{\perp}_{xyzw}$ extracted from the quantum metric tensor and Berry curvature in Fig. 2-9, showing the characteristic inverse-cube dependence on the radial coordinate. Again, the errorbars here are propagated from fitting errors of resonant frequencies and Rabi oscillations.

Now, the measurement of the quantized topological charge ($\mathcal{DD}$ invariant), together with the characteristic radial behavior of a monopole field, fully confirm the existence of a tensor monopole in our synthetic 4D parameter space.

We remark that our precise control over the Weyl-type Hamiltonian illustrates the potential offered by solid-state qudits in the realm of quantum simulation. Interesting perspectives include the fate of tensor monopoles upon coupling the system to other spins or qubits [128], and the study of non-Abelian structures induced by spectral degeneracies and tensor fields [135]. Also, our method could be applied to higher dimensions, where higher-order tensor gauge fields exist and the demonstrated tensor monopole is naturally generalized.

Furthermore, we note that Eq. 3.4 suggests that the physics of tensor monopoles could be investigated in systems of particles moving on a 4D lattice, where $\mathbf{q}$ would represent the corresponding crystal momenta. Such 4D Weyl lattice systems have been recently proposed [132, 136] and could be realized in quantum-engineered systems, extending the 3D lattice where particles lie with a synthetic dimension [137, 138].

3.1.4 Topological spectral transition triggered by external field

Symmetry analysis

As mentioned, the Hamiltonian Eq. 3.4 studied above preserves the chiral symmetry and two mirror symmetries. We now break the chiral symmetry by adding a longitudinal field to the Weyl-type Hamiltonian

$$\mathcal{H}_{ST} = \mathcal{H}_{4D} + diag(B_z, 0, -B_z)/\sqrt{2}. \tag{3.22}$$

and study its spectral properties. The field is realized by detuning the dual-frequency microwave driving by equal and opposite amounts, as it's clearly shown in Eq. 2.18.

The new Hamiltonian $\mathcal{H}_{ST}$ still preserves the two mirror symmetries,

$$\begin{aligned}
M_1\mathcal{H}_{ST}(q_x, q_y, q_z, q_w)M_1^{-1} &= \mathcal{H}_{ST}(-q_x, -q_y, q_z, q_w), \\
M_2\mathcal{H}_{ST}(q_x, q_y, q_z, q_w)M_2^{-1} &= \mathcal{H}_{ST}(q_x, q_y, -q_z, -q_w),
\end{aligned} \tag{3.23}$$

with $M_1 = diag(-1, 1, 1)$ and $M_2 = diag(1, 1, -1)$, keeping it gapless. These two mirror symmetries naturally imply the inversion symmetry:

$$U_I\hat{H}_{ST}(q_x, q_y, q_z, q_w)U_I^{-1} = \hat{H}_{ST}(-q_x, -q_y, -q_z, -q_w) \tag{3.24}$$

where $U_I = M_1 M_2$. The subspaces of model Eq. 3.22 can further possess PT symmetries [114].

Upon application of the field, the system undergoes a topological spectral transition from the 4D Weyl-like structure to a new symmetry-protected energy spectrum that features a pair of doubly degenerate surfaces in the $\beta - \phi$ phase (with $\alpha = 0(\pi/2), B_z = H_0$). In cartesian coordinate (q_x, q_y, q_z, q_w), the degenerate surfaces correspond to two spectral rings in the $q_x - q_y$ ($q_z - q_w$) space along $q_z = q_w = 0$ ($q_x = q_y = 0$) with radius B_z. For example, the eigenvalue analytical forms in the subspace of $q_z = q_w = 0$ are

$$\epsilon_- = \frac{1}{2}\left(\frac{B_z}{\sqrt{2}} - \sqrt{\frac{B_z^2}{2} + 4q_x^2 + 4q_y^2}\right),$$

$$\epsilon_0 = -\frac{B_z}{\sqrt{2}}, \tag{3.25}$$

$$\epsilon_+ = \frac{1}{2}\left(\frac{B_z}{\sqrt{2}} + \sqrt{\frac{B_z^2}{2} + 4q_x^2 + 4q_y^2}\right),$$

and the lower two bands become degenerate when $q_x^2 + q_y^2 = B_z^2$. This corresponds to the case when $\alpha = 0$. At $\alpha = \pi/2$, there is another nodal (or spectral) ring between the middle and upper bands. In short, the two nodal rings occur at

$$q_x^2 + q_y^2 = B_z^2, \quad q_z = q_w = 0,$$
$$\text{or} \quad q_z^2 + q_w^2 = B_z^2, \quad q_x = q_y = 0. \tag{3.26}$$

We show the typical spectrum in Fig. 3-3.

Experiment observables

To identify the signatures of the spectral rings, we use two observables inspired by the tensor-monopole measurements

$$\mathcal{G} = 8\int \epsilon_{\mu\nu\lambda}\sqrt{\det g_{\bar{\mu}\bar{\nu}}}\, d\alpha$$
$$\mathcal{B} = -\int (\mathcal{F}_{\alpha\beta} + \mathcal{F}_{\phi\alpha})\, d\alpha. \tag{3.27}$$

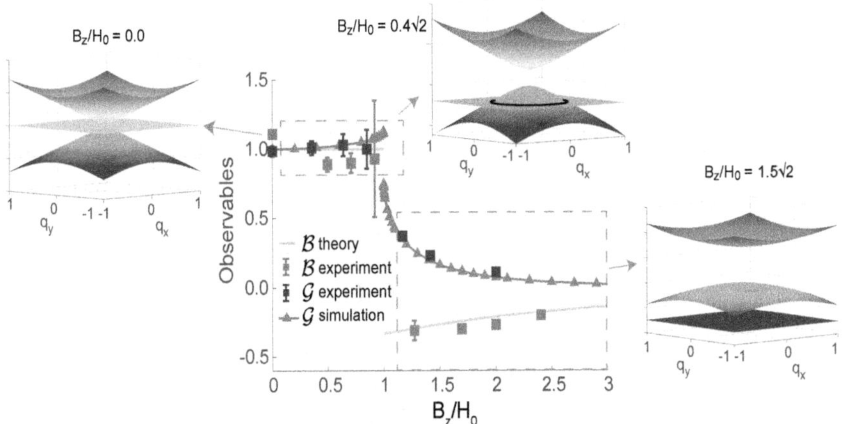

Figure 3-3:   The central plot shows experimental data (blue squares) and numerical simulation (green triangles) of the experimental observable $\mathcal{G}$ based on the metric tensor, experimental data (red squares) and analytical result (yellow line) of the observable $\mathcal{B}$ based on the Berry curvature. Both methods shows a sharp response at $B_z = H_0$ when the spectral rings cross the boundary of the integration hypersphere. The experimental observable $\mathcal{G}$, $\mathcal{B}$ correspond to the $\mathcal{DD}$ invariant when $B_z = 0$ and chiral symmetry is preserved. On the side we show 3 representative energy spectra as the longitudinal field B_z increases ($q_z = q_w = 0$). The external field splits the triply degenerate Weyl node (left) into doubly degenerate spectral rings (middle). As the field further increases, the system becomes gapped in the enclosed integration hypersphere (right).

They represent integration over a hyperspherical surface with radius H_0, and correspond to the $\mathcal{DD}$ invariant when $B_z = 0$.

Indeed, with the parameterization given in Eq. 2.18, the observable $\mathcal{B}$ in the presence of external field ($B_z \neq 0$) can be analytically calculated, as shown below. By explicitly evaluating the integral of Eq. 3.12, we have

$$\mathcal{B} = \frac{1}{2\pi^2} \int_{S_3} \mathcal{H}_{\alpha\beta\phi} d\alpha d\beta d\phi = [v_1^2(0) - v_3^2(0)] - [v_1^2(\pi/2) - v_3^2(\pi/2)] \tag{3.28}$$

and the Hamiltonian eigenvectors can be calculated easily for $\alpha = 0, \pi/2$. By setting

$h = B_z/H_0$, we have

$$v_1(0) = \begin{cases} -\sqrt{\frac{1}{2}\left(1 - \frac{h}{\sqrt{h^2+8}}\right)}, & h < 1 \\ 0 & \text{otherwise} \end{cases} \quad , v_3(0) = \begin{cases} 0, & h < 1 \\ 1 & \text{otherwise} \end{cases}$$

$$\text{(3.29)}$$

$$v_1(\pi/2) = 0 \qquad v_3(\pi/2) = \sqrt{\frac{1}{2}\left(1 + \frac{h}{\sqrt{h^2+8}}\right)}$$

Now it's clear to see that the analytical form for $\mathcal{B}$

$$\mathcal{B} = \begin{cases} 1, & B_z < H_0 \\ -\frac{1}{2}\left(1 - \frac{B_z}{\sqrt{B_z^2+8H_0^2}}\right) & \text{otherwise} \end{cases}, \qquad \text{(3.30)}$$

which can match well with experiments.

As the field strength B_z increases, the two nodal rings expand away from the origin. When $B_z < H_0$, they are enclosed in the integration hypersphere (as shown in spectrum in Fig. 3-3). The nodal rings then cross the boundary of our integration hypersphere at $B_z = H_0$. When $B_z > H_0$, the spectrum becomes gapped inside the integration hypersphere and the nodal rings are outside. For various B_z, we perform linear and elliptical parametric modulators to reconstruct the metric tensor and the Berry curvature (detailed measurement results in the supporting material of Ref [114]), from which we obtain $\mathcal{G}$, $\mathcal{B}$. Remarkably, both experimental observables clearly signal the nodal rings crossing the hypersphere at $B_z = H_0$, an indication of the topological nodal ring semimetal phase. The experimental results are in agreement with the simulation for $\mathcal{G}$.

These results show that, as B_z increases while keeping H_0 fixed (i.e. restricting ourselves to a hypersphere in parameter space), our system simulates phase transitions from the Weyl-type nodes ($B_z = 0$), to the topological spectral rings ($B_z < H_0$) characterized by a constant $\mathcal{B}$, and eventually to a gapped spectrum ($B_z > H_0$).

Insights on unconventional quasiparticles

From the viewpoint of topological protected phases and novel quantum matters, our synthetic Hamiltonian could serve as a playground for exploring unconventional quasi-particles beyond Dirac and Weyl fermions in high dimensional space [139, 140]. Recall that our Hamiltonian is given by Eq. 2.18. The linearized $\mathbf{k} \cdot \mathbf{p}$ Hamiltonian for space group (SG) 220 is given by:

$$\mathcal{H}_{220}(\mathbf{k}) = \begin{pmatrix} 0 & k_y & k_x \\ k_y & 0 & -k_z \\ k_x & -k_z & 0 \end{pmatrix}, \tag{3.31}$$

We note that when we take a slice of our Hamiltonian

$$\mathbf{k} = (k_x, k_y, k_z) = (H_0 \sin(\alpha - \pi/4), H_0 \cos(\alpha - \pi/4), B_z/\sqrt{2}), \tag{3.32}$$

Eq. 2.18 and Eq. 3.31 have identical eigenvalue spectrums for any β, ϕ.

For the Hamiltonian in Eq. 3.31, pairs of two bands are degenerate along $|k_x| = |k_y| = |k_z|$. This corresponds to the condition $B_z = H_0$ and $\alpha = 0, \pi/2$ in our model, matching our numerical simulation and experimental results. Due to the rotation symmetry of β and ϕ in our model, we remark that a pair of doubly degenerate nodal surfaces along $(\alpha = 0, \pi/2, B_z = H_0)$ emerge when $B_z \neq 0$ in our 4D parameter space. This corresponds to the two nodal rings in the (q_x, q_y, q_z, q_w) coordinate as discussed above.

In short, the equivalence between the SG220 model and a slice of our model presented here implies that the doubly degeneracy induced by the detuning B_z has a correspondence to crystal symmetry-protected fermionic excitations, and the observed phase transition has a non-trivial topological meaning. It would be interesting to use our experimental system to further simulate the topological properties of the SG220 model (and other triple-degenerate point models).

3.1.5 Topological phase transition triggered by manifold displacement

In the above discussions we introduced a topological spectral transition that was triggered by the external field $diag(B_z, 0, -B_z)/\sqrt{2}$. Indeed, the most straightforward topological phase transition could be induced by displacement of the integration hypersphere along one of the parameter axis , e.g., $q_x \to q_x + \delta_x$. The Hamiltonian can be written as:

$$\hat{H}_{disp} = \begin{pmatrix} 0 & H_0 \cos \alpha e^{-i\beta} + \delta_x & 0 \\ H_0 \cos \alpha e^{i\beta} + \delta_x & 0 & H_0 \sin \alpha e^{i\phi} \\ 0 & H_0 \sin \alpha e^{-i\phi} & 0 \end{pmatrix}. \tag{3.33}$$

We can then analytically calculate the 3-form Berry curvature in the presence of displacement:

$$\mathcal{H}_{\alpha\beta\phi}(H_0, \alpha, \beta, \delta_x) = \frac{H_0^3 \cos \alpha \sin \alpha (H_0 + \delta_x \cos \alpha \cos \beta)}{(\delta_x^2 + H_0^2 + 2H_0\delta_x \cos \alpha \cos \beta)^2} \tag{3.34}$$

and evaluate its integral to find the $\mathcal{DD}$ invariant. The topological phase transition is characterized by the $\mathcal{DD}$ invariant, as shown in Fig. 3-4, where $\mathcal{DD} = 1 \to 0$ when $|\delta_x/H_0| > 1$. Similar experiment has been done in superconducting qubit [134].

In our experiments, we chose to add the fictitious z field to the system instead of this displacement to preserve rotation symmetry about β, ϕ which can greatly reduce the total measurement time. Moreover, the case of adding a fictitious z field has more intriguing physics as discussed above.

3.2 Connection with quantum multi-parameter estimation

In this section, we consider another application of QGT: exploring geometry in quantum multi-parameter estimation problems [78]. We will discuss precision limits of

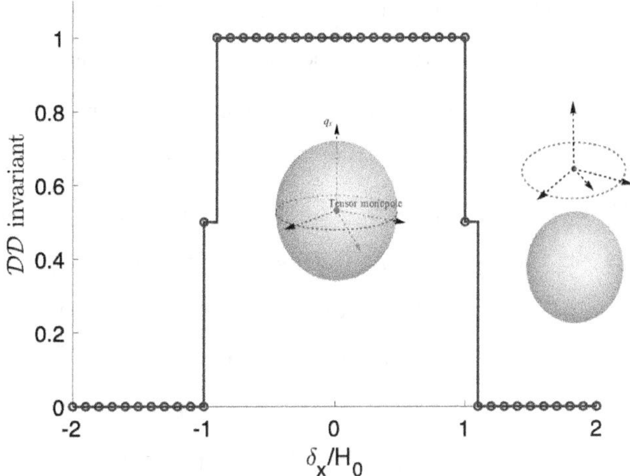

Figure 3-4:   Phase transition triggered by a manifold displacement. When $|\delta_x/H_0| <$ 1, the manifold encloses the triply degenerate point and $\mathcal{DD} = 1$; when $|\delta_x/H_0| > 1$, the monopole is no longer enclosed, leading to a trivial phase $\mathcal{DD} = 0$.

multi-parameter estimation through the lens of quantum geometry and based on the above results, we will investigate and experimentally extract the attainable precision bound for three-parameter estimations for a given quantum state.

3.2.1   Relating quantum geometry to multi-parameter sensing

Overview of the problem

In quantum metrology, the precision limit is usually quantified by the quantum Cramér-Rao bound (QCRB) [141, 142, 143, 144, 145, 146, 147]. For unbiased estimation of an unknown system parameter, the QCRB is given by the inverse of the quantum Fisher information [144, 145, 146, 147]. The estimator is chosen to be optimal, therefore the QCRB only depends on the quantum state. In turn, the precision of estimating a parameter can be linked to the "distance" between two nearby states differing by an infinitesimally small parameter change [148, 149, 112, 146]. This naturally connects quantum sensing to the quantum geometric properties of the system, as the real part of the QGT, quantum metric tensor, is closely related to the ultimate estimation precision quantified by the quantum Fisher information.

However, this picture becomes more involved when extending the goal from estimating a single parameter to multiple parameters. The complication arises from the incompatibility between each parameter's optimal estimators–a signature of quantum mechanics [147, 146, 149]. This leads to trade-offs among the estimation precisions of different parameters when picking the initial state. This interplay among parameters makes the multi-parameter estimation problem more intriguing than single parameter estimation. In the following, we show that this incompatibility, quantified by the non-commutativity of optimal measurement operators for different parameters, can be captured by the imaginary part of QGT for a given quantum state [145, 146, 147].

Therefore, by experimentally characterizing QGT of quantum states, we will be able to explore the problem of quantum multi-parameter estimation and evaluate the attainable QCRB from a geometric perspective.

Introduction on quantum Fisher information and Cramér-Rao bound

To begin with, we consider a generic quantum multi-parameter estimation problem given by a quantum statistic model ρ_θ, a family of density operator labelled by $\boldsymbol{\theta} = (\theta_1, \theta_2, ..., \theta_d)^T$ that is the set of unknown parameters to be estimated. Performing POVM $\mathcal{M} = \{\mathcal{M}_k | \sum_k \mathcal{M}_k = I, \mathcal{M}_k \geq 0\}$, we obtain the measurement conditional probabilities $p(k|\boldsymbol{\theta}) = \text{Tr}[\rho_\theta \mathcal{M}_k]$. With N repeated measurements, the unknown parameters are estimated via the estimator $\tilde{\boldsymbol{\theta}}(k)$. The accuracy of the estimation is quantified by the mean-square error matrix $\boldsymbol{V}(\boldsymbol{\theta}) = \sum_k p(k|\boldsymbol{\theta})[\tilde{\boldsymbol{\theta}}(k) - \boldsymbol{\theta}][\tilde{\boldsymbol{\theta}}(k) - \boldsymbol{\theta}]^T$. Obtaining the lower bounds of this error matrix is of critical importance in quantum sensing and metrology.

The bounds usually depend on the choice of POVM. The symmetric logarithmic derivative [141, 142, 143] (SLD) can be used to define a bound that only depends on the quantum statistic model ρ_θ, having optimized over the measurement operators. The SLD L_μ ($\mu \in \boldsymbol{\theta}$) is implicitly defined via the equation $\partial_\mu \rho_\theta = (L_\mu \rho_\theta + \rho_\theta L_\mu)/2$. The corresponding quantum Fisher information matrix (QFIM), $\boldsymbol{J}(\boldsymbol{\theta})$, is given by $J_{\mu\nu}(\boldsymbol{\theta}) = \text{Tr}[\rho_\theta (L_\mu L_\nu + L_\nu L_\mu)/2]$, which yields the matrix SLD QCRB $\boldsymbol{V}(\boldsymbol{\theta}) \geq \boldsymbol{J}(\boldsymbol{\theta})^{-1}$.

It is convenient to introduce a scalar QCRB, $C^S(\boldsymbol{\theta}, \boldsymbol{W}) = \text{tr}\{\boldsymbol{W}\boldsymbol{J}(\boldsymbol{\theta})^{-1}\}$. For a weight matrix $\boldsymbol{W}$, the inequality becomes $\text{tr}\{\boldsymbol{W}\boldsymbol{V}\} \geq C^S(\boldsymbol{\theta}, \boldsymbol{W})$. This SLD-CRB is generally not attainable due to the incompatibility of generators of different parameters. That is, the optimal measurement operators corresponding to different parameters do not commute with each other, making this scalar bound unreachable.

Holevo derived [144] a tighter scalar bound, the Holevo Cramér-Rao bound (HCRB) $C^H(\boldsymbol{\theta}, \boldsymbol{W})$, which is however only obtained as an optimization. More recently, it has been shown that these two scalar bounds satisfy the following inequality [145]

$$\begin{aligned}
C^S(\boldsymbol{\theta}, \boldsymbol{W}) &\leq C^H(\boldsymbol{\theta}, \boldsymbol{W}) \\
&\leq C^S(\boldsymbol{\theta}, \boldsymbol{W}) + ||\sqrt{\boldsymbol{W}}\boldsymbol{J}^{-1}\boldsymbol{F}\boldsymbol{J}^{-1}\sqrt{\boldsymbol{W}}||_1 \\
&\leq (1+\gamma)C^S(\boldsymbol{\theta}, \boldsymbol{W}) \leq 2C^S(\boldsymbol{\theta}, \boldsymbol{W}),
\end{aligned} \tag{3.35}$$

where $|| \cdot ||_1$ denotes the trace norm and $\boldsymbol{F}$ is known as the mean Uhlmann curvature with element $F_{\mu\nu} = -\frac{i}{4}\text{Tr}[\rho_{\boldsymbol{\theta}}[L_\mu, L_\nu]]$ (for pure states this reduces to the Berry curvature.) The *characterization number* γ that appears in Eq. (3.35) is defined as

$$\gamma = ||i2\boldsymbol{J}^{-1}\boldsymbol{F}||_\infty, \tag{3.36}$$

where $||\boldsymbol{A}||_\infty$ denotes the largest eigenvalue of $\boldsymbol{A}$. γ characterizes the "quantumness" of the system, i.e., it quantifies the amount of incompatibility of the statistic model $\rho_{\boldsymbol{\theta}}$. As implicit in Eq. (3.35) and explained hereafter, it can be proved that $0 \leq \gamma \leq 1$ [145].

An explicit expression for an attainable scalar QCRB with weight matrix $\boldsymbol{W}=\boldsymbol{J}$, which monotonically increases with γ, was derived in Ref. [150, 151]

$$\begin{aligned}
C(\boldsymbol{\theta}) &= \text{tr}\left\{ \text{Re}\left[\sqrt{d + 2i\boldsymbol{J}(\boldsymbol{\theta})^{-1/2}\boldsymbol{F}(\boldsymbol{\theta})\boldsymbol{J}(\boldsymbol{\theta})^{-1/2}} \right]^{-2} \right\} \\
&= \sum_i \frac{2}{1 + \sqrt{1 - |\gamma_i|^2}},
\end{aligned} \tag{3.37}$$

where d is the d-dimensional identity matrix and γ_i is the i-th eigenvalue of

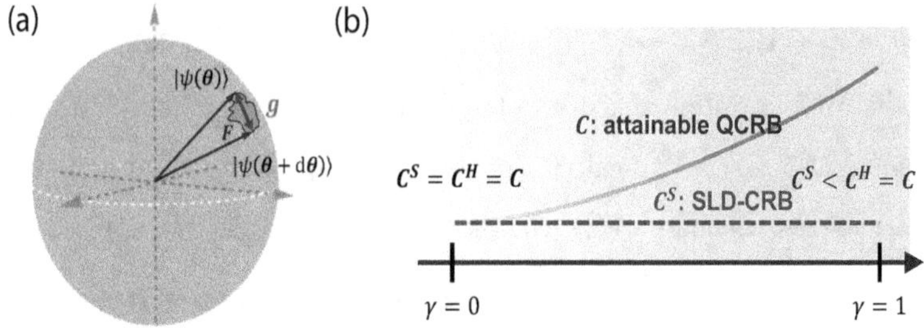

Figure 3-5: (a). Diagram showing the "distance" between nearby states $|\psi(\boldsymbol{\theta})\rangle$ and $|\psi(\boldsymbol{\theta}+d\boldsymbol{\theta})\rangle$ in parameter space, from which the metric tensor $\boldsymbol{g}$ (proportional to quantum Fisher information $\boldsymbol{J}$) and Berry curvature $\boldsymbol{F}$ naturally arise. (b). SLD-CRB and attaianable QCRB for different γ values. At $\gamma = 0$ the model is quasi-classical and the SLD-CRB is reachable, i.e., $C^S = C^H = C$, while the model is called coherent at $\gamma = 1$ with $C^S < C^H = C$. Here we take the weight matrix $\boldsymbol{W} = \boldsymbol{J}$ and denote $C^{S(H)}(\boldsymbol{\theta}, \boldsymbol{W})$ as $C^{S(H)}$ for simplicity.

$i2\boldsymbol{J}^{-1}\boldsymbol{F}$. In Fig. 3-5 (b), we show the relation of the scalar QCRBs as a function of γ: at $\gamma = 0$ the model is quasi-classical and we have $C^S = C^H = C$; while in the other extreme case, at $\gamma = 1$ the model is coherent (the most quantum case) with $C^S < C^H = C$.

Relation between QGT and multiparameter estimation

Now, to show the relation between QGT and parameter estimation, one can assume the parameter $\boldsymbol{\theta}$ to be introduced by evolving an initial state $|\psi_0\rangle$ via a unitary evolution, i.e., $|\psi(\boldsymbol{\theta})\rangle = U(\boldsymbol{\theta})|\psi_0\rangle$. Then, the generator for parameter $\mu \in \boldsymbol{\theta}$ is the operator $\mathcal{G}_\mu \equiv i(\partial_\mu U)U^\dagger$ that describes the derivative of the unitary operator:

$$|\psi(\boldsymbol{\theta})\rangle = \exp[-iH(\boldsymbol{\theta})t]|\psi_0\rangle = U(\boldsymbol{\theta})|\psi_0\rangle$$
$$\rightarrow U(\boldsymbol{\theta}+d\theta_j)|\psi_0\rangle = [I + (\partial_{\theta_j}U)U^\dagger d\theta_j]U|\psi_0\rangle \qquad (3.38)$$
$$= \exp\{-i[i(\partial_{\theta_j}U)U^\dagger]d\theta_j\}U|\psi_0\rangle.$$

The generators would help us calculate the geometric quantities, as we show now.

To begin with, we can apply $|\psi(\boldsymbol{\theta})\rangle = U(\boldsymbol{\theta})|\psi_0\rangle$ to the definition of the QGT:

$$
\begin{aligned}
\chi_{\mu\nu} &= \langle\partial_\mu\psi|\,(I - |\psi\rangle\langle\psi|)\,|\partial_\nu\psi\rangle \\
&\rightarrow \langle\partial_\mu(U\psi_0)|\,(I - U|\psi_0\rangle\langle\psi_0|U^\dagger)\,|\partial_\nu(U\psi)\rangle \\
&= \langle\psi_0|\,\partial_\mu U^\dagger\partial_\nu U\,|\psi_0\rangle - \langle\psi_0|\,(\partial_\mu U^\dagger)U\,|\psi_0\rangle\langle\psi_0|\,(U^\dagger\partial_\nu U)\,|\psi_0\rangle
\end{aligned}
\tag{3.39}
$$

Then, the real part of QGT $\chi_{\mu\nu}$ is given by:

$$
\begin{aligned}
g_{\mu\nu} &= \frac{1}{2}(\chi_{\mu\nu} + \chi_{\mu\nu}^\dagger) \\
&= \frac{1}{2}\langle\psi_0|\,(\partial_\mu U^\dagger\partial_\nu U + \partial_\nu U^\dagger\partial_\mu U)\,|\psi_0\rangle \\
&\quad + \langle\psi_0|\,(\partial_\mu U^\dagger)U\,|\psi_0\rangle\langle\psi_0|\,(\partial_\nu U^\dagger)U\,|\psi_0\rangle\,.
\end{aligned}
\tag{3.40}
$$

Because $\mathcal{G}_\mu$ is Hermitian, $\mathcal{G}_\mu = \mathcal{G}_\mu^\dagger$, meaning $i(\partial_\mu U)U^\dagger = -iU(\partial_\mu U^\dagger)$. Using this equality, Eq. 3.40 now becomes

$$
\begin{aligned}
g_{\mu\nu} &= \frac{1}{2}\langle\psi_0|\,((\partial_\mu U^\dagger)UU^\dagger\partial_\nu U + (\partial_\nu U^\dagger)UU^\dagger(\partial_\mu U))\,|\psi_0\rangle \\
&\quad - \langle\psi_0|\,(i\partial_\mu U^\dagger)U\,|\psi_0\rangle\langle\psi_0|\,(i\partial_\nu U^\dagger)U\,|\psi_0\rangle \\
&= \frac{1}{2}\langle\psi_0|\,\{\mathcal{G}_\mu,\mathcal{G}_\nu\}\,|\psi_0\rangle - \langle\psi_0|\,\mathcal{G}_\mu\,|\psi_0\rangle\langle\psi_0|\,\mathcal{G}_\nu\,|\psi_0\rangle \\
&= \mathrm{cov}(\mathcal{G}_\mu,\mathcal{G}_\nu) = \frac{1}{4}J_{\mu\nu},
\end{aligned}
\tag{3.41}
$$

where $\mathrm{Cov}(\mathcal{G}_\mu,\mathcal{G}_\nu)$ is the covariance of generators.

That is, the QFIM element $J_{\mu\nu}$ and the metric tensor $g_{\mu\nu}$ are related by

$$
J_{\mu\mu} = 4\left\langle\Delta\mathcal{G}_\mu^2\right\rangle = 4g_{\mu\mu}, \quad J_{\mu\nu} = 4\mathrm{Cov}(\mathcal{G}_\mu,\mathcal{G}_\nu) = 4g_{\mu\nu},
\tag{3.42}
$$

where $\left\langle\Delta\mathcal{G}_\mu^2\right\rangle$ is the generator variance and $\mathrm{Cov}(\mathcal{G}_\mu,\mathcal{G}_\nu)$ their covariance. This relation is intuitive since the metric tensor indeed characterizes the overlap between the nearby states. A larger $g_{\mu\nu}$ indicates a larger distinguishability between two states differing by an infinitesimal change of parameters, and it is thus related to a larger quantum Fisher information.

Similarly, the Berry curvature is given by the commutator between the parameter generators

$$F_{\mu\nu} = i(\chi_{\mu\nu} - \chi_{\mu\nu}^{\dagger})$$
$$= i \langle\psi_0| \left((\partial_\mu U^\dagger)UU^\dagger\partial_\nu U - (\partial_\nu U^\dagger)UU^\dagger(\partial_\mu U)\right) |\psi_0\rangle \tag{3.43}$$
$$= i \langle\psi_0| [\mathcal{G}_\mu, \mathcal{G}_\nu] |\psi_0\rangle .$$

That is,

$$F_{\mu\nu} = i \langle[\mathcal{G}_\mu, \mathcal{G}_\nu]\rangle . \tag{3.44}$$

It is now clear that the Berry curvature matrix characterizes the incompatibility of the system as it arises from the non-commutativity between generators of different parameters.

Characterization number and uncertainty relation

We next show that in the opposite limit the characterization number γ is upper-bounded by the fundamental quantum uncertainty principles. We consider the two-parameter case and start from the Robertson-Schrödinger uncertainty relation for two operators $\hat{A}, \hat{B}$ [152], the stronger version of the better-known Heisenberg uncertainty relation. In the quantum parameter estimation case, the operators $\hat{A}, \hat{B}$ can be replaced by the generators of parameters μ, ν, giving

$$\langle\Delta\mathcal{G}_\mu^2\rangle \langle\Delta\mathcal{G}_\nu^2\rangle \geq \frac{1}{4}| \langle[\mathcal{G}_\mu, \mathcal{G}_\nu]\rangle |^2 + \mathrm{Cov}(\mathcal{G}_\mu, \mathcal{G}_\nu)^2. \tag{3.45}$$

With Eq. (3.36,3.42,3.44) we have

$$\gamma = 2\frac{F_{\mu\nu}}{\sqrt{\det(\boldsymbol{J})}} = \frac{1}{2}\frac{F_{\mu\nu}}{\sqrt{\det(\boldsymbol{g})}} \leq 1, \tag{3.46}$$

which relates γ to half the Berry's phase along the curve that encloses a unit area induced by metric tensor. This relation holds for any two-parameter space in a multi-parameter setting. The above relation can also be derived from the fact that the QGT $\chi_{\mu\nu}$ is a positive semi-definite matrix and has been explored in various models [153, 154, 155].

Reversely, generalization of the two-operator product uncertainty relations Eq. 3.45 to multi-operators can be performed using the fact that $\gamma \leq 1$. For example, we consider three operators $\hat{A}$, $\hat{B}$ and $\hat{C}$ that satisfies $Cov(\hat{A}, \hat{B}) = Cov(\hat{A}, \hat{C}) = 0$ and $[\hat{B}, \hat{C}] = 0$ (this is indeed the case for the three generators in the 3-parameter state $|u_-\rangle$ studied above). Then, the following equality can be deduced from $\gamma \leq 1$:

$$\begin{aligned}
\left\langle \Delta\hat{A}^2 \right\rangle \left\langle \Delta\hat{B}^2 \right\rangle \left\langle \Delta\hat{C}^2 \right\rangle \geq {}& \left\langle \Delta\hat{A}^2 \right\rangle Cov(\hat{B}, \hat{C})^2 \\
& + \frac{1}{4}(\left\langle \Delta\hat{C}^2 \right\rangle | \left\langle [\hat{A}, \hat{B}] \right\rangle |^2 + \left\langle \Delta\hat{B}^2 \right\rangle | \left\langle [\hat{A}, \hat{C}] \right\rangle |^2 \\
& + 2Cov(\hat{B}, \hat{C}) | \left\langle [\hat{A}, \hat{B}] \right\rangle || \left\langle [\hat{A}, \hat{C}] \right\rangle |).
\end{aligned} \tag{3.47}$$

More generally, one can have the quantum Fisher information matrix in a simple diagonal form under certain parameter transformations. The vanished off-diagonal terms then correspond to $Cov(\hat{X}, \hat{Y}) = 0$ for $\forall \hat{X}, \hat{Y} \in \{\hat{A}, \hat{B}, \hat{C}\}$. Again, from the fact that $\gamma = ||i2\boldsymbol{J}^{-1}\boldsymbol{F}||_\infty \leq 1$, we reach the generalized three-operator Heisenberg uncertainty relation

$$\begin{aligned}
\left\langle \Delta\hat{A}^2 \right\rangle \left\langle \Delta\hat{B}^2 \right\rangle \left\langle \Delta\hat{C}^2 \right\rangle \geq {}& \frac{1}{4}(\left\langle \Delta\hat{C}^2 \right\rangle | \left\langle [\hat{A}, \hat{B}] \right\rangle |^2 \\
& + \left\langle \Delta\hat{B}^2 \right\rangle | \left\langle [\hat{A}, \hat{C}] \right\rangle |^2 + \left\langle \Delta\hat{A}^2 \right\rangle | \left\langle [\hat{B}, \hat{C}] \right\rangle |^2)
\end{aligned} \tag{3.48}$$

The above product inequality has also been derived by constructing positive semi-definite matrices [156, 157, 158]. The product-form uncertainty relation for N operators can be constructed in a similar manner. To this end, we have shown the connection between quantum geometry with the generalized uncertainty principles.

3.2.2 Experimental extraction of attainable bound

Following the QGT measurements in Chapter 2, we next consider a specific case of multi-parameter estimation problem and experimentally extract the attainable quantum Cramér-Rao bound.

3-parameter state $|u_-\rangle$

With our previous measurement of QGT of the state $|u_-\rangle$ (Eq. 2.21) in Fig. 2-9, we can extract the quantum Fisher information and Berry curvature matrix and then get the characterization number γ according to Eq. 3.36. In Fig. 3-6, we show the experimental values of γ as well as the attainable QCRB. When $\alpha = 0$ or $\frac{\pi}{2}$, the

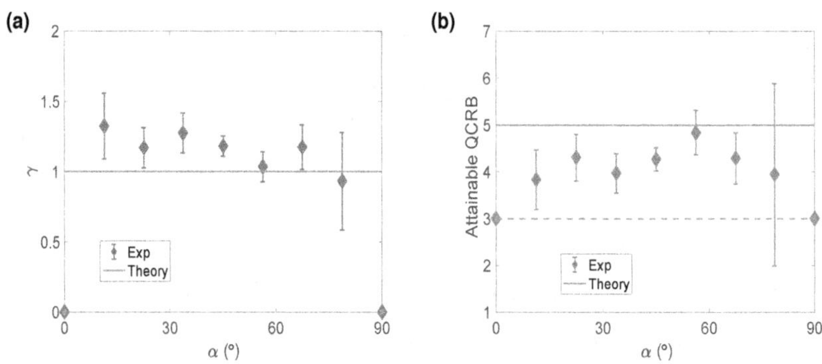

Figure 3-6:   (a) Characterization number γ and (b) the corresponding attainable QCRB for the three-parameter model. The solid lines are theoretical predictions in both plots. The dashed line in (b) is the attainable QCRB when $\gamma = 0$, i.e., the SLD-CRB for the model. The errorbars are propagated from the errors in quantum geometric quantities.

parameterized model Eq. 2.21 effectively becomes a qubit system in a one-parameter (sub)space. For example, when $\alpha = 0$ we have $J_{\phi\phi} = 0$, meaning that we have no information on ϕ, and $\boldsymbol{F} = 0$, indicating that there are no non-commuting generators, and the parameter β can be estimated trivially. The model is thus quasi-classical. In this case, the attainable QCRB is equal to the SLD-CRB and HCRB, i.e., $C = C^S = C^H$ due to the quasi-classical nature of the system.

When $\alpha \in (0, \frac{\pi}{2})$, however, the model is coherent with $\gamma = 1$, indicating the maximal non-commutativity of the generators corresponding to the three parameters. Now the attainable QCRB is equal to the HCRB [151] and is higher than SLD-CRB due to the incompatibility of the three parameters.

It's interesting to remark that for this particular quantum state $|u_-\rangle$ here, one can rewrite γ as

$$\gamma = 2\frac{\sqrt{J_{\phi\phi}F_{\alpha\beta}^2 + J_{\beta\beta}F_{\alpha\phi}^2 - 2J_{\beta\phi}F_{\alpha\beta}F_{\alpha\phi}}}{\sqrt{\det \boldsymbol{J}}} \equiv \frac{2\mathcal{H}_{\alpha\beta\phi}}{\sqrt{\det \boldsymbol{J}}}. \tag{3.49}$$

When $\gamma = 1$, one recovers the relation between the generalized 3-from curvature $\mathcal{H}$ and metric tensor $\boldsymbol{g}$, as in Eq. 3.6. Indeed, more generally γ can be expressed as the ratio between an $n-$form generalized Berry curvature $F_{\alpha\beta...}$ and the determinant of the metric tensor, $\gamma = c\frac{F_{\alpha\beta...}}{\sqrt{det(g)}}$ with c being a constant normalization factor. Accordingly, γ is gauge-invariant and for pointlike or extended monopoles [131, 114], we have $\gamma(\boldsymbol{\theta}) = 1$, $\forall\boldsymbol{\theta}$ in parameter space, indicating the largest quantumness of these monopoles. The isotropy of γ is due to the radial nature of the field emanating from monopole sources; when $\gamma = 1$, the flux (i.e., the generalized Berry curvature) through an unit area (given by the metric tensor determinant) is unit and thus the flux integral over a closed sphere S^n gives the quantized topological number, upon normalization.

2-parameter subspace of a 3-parameter state

Above we have shown the cases of $\gamma = 0, 1$. To study intermediate cases where $\gamma \leq 1$, one can consider different geometrical models [159] or, as done here, rely on the two-parameter subspaces of the above three-parameter model where γ is a function of α. From the viewpoint of the Hamiltonian, this corresponds to taking a slice of the original gapless Weyl-type system, which yields a gapped topological insulator that can have $\gamma \leq 1$.

We thus focus on the subspace spanned by parameters (α, β) and (α, ϕ). In Fig. 3-7 we plot the characterization number γ as well as the corresponding attainable QCRB. When $\alpha = 0(\frac{\pi}{2})$, the three-level system effectively reduces to a two-level system and the parameter $\phi(\beta)$ cannot be estimated due to the null population in $[0, 0, 1]^T$ ($[1, 0, 0]^T$). We observe γ varies continuously from 0 to 1 when sweeping α in the two subspaces. The HCRB thus ranges from 2, equal to the scalar SLD-CRB, (asymptotically) to 4, where the attainable scalar QCRB is largest (two times larger

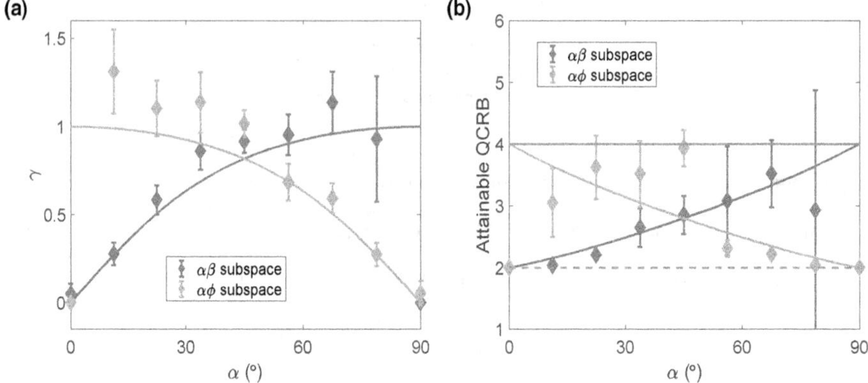

Figure 3-7: .   Characterization number **(a)** and attainable QCRB **(b)** for two-parameter estimation in the subspace of three-parameter model. The solid colored lines are theoretical calculations in both figures. At $\alpha = 0(\pi/2)$, the predicted value for γ (attainable QCRB) is 0 (2). The solid (dashed) black lines in (b) are attainable QCRB values when $\gamma = 1$ ($\gamma = 0$).

than the SLD-CRB, as bounded by Eq. 3.35). These results show how γ directly predicts the attainable QCRB for subspace models, and thus sets the estimation precision when evaluating a subset of parameters. The analysis above can then be applied to scenarios where there are nuisance parameters, i.e., parameters that are not of interest but nevertheless affect the precision of estimating other parameters of interest [160, 161, 162].

Chapter 4

NV center as quantum sensors

As shown in the previous chapter, sensing multi-parameters is usually difficult. We now focus on sensing one specific parameter or signal of interest. As discussed in Chapter 1, the NV community has witnessed rapid development in designing NV-based temperature sensors [39, 40, 41], magnetic fields sensors [163, 65] and so on. In particular, NV centers in nanodiamond (ND) have been employed in sensing chemical and biological processes. The system has superior properties [26, 62], for example, they possess very high photo-stability and high thermal stability without significant photo-bleaching effect at room temperature. Hence they can be utilized as fluorescent makers to track molecules in biological cells or tissues [79, 80, 62, 63]. The size of ND with NV center inside can be fabricated down to several nanometers, which provide excellent spatial resolution in sensing. For example, small NDs can be mounted on the tip of atomic force microscopy to measure local environments with resolution in nanometers [89, 90]. Furthermore, the fact that ND is non-toxic makes it preferable compared with many organic dyes. It has been shown that NDs are exceptionally biocompatible in cells and can be easily taken up through different pathways [164, 165, 166]. Functional NDs with appropriate surface modulation and chemical conjugation could further enable multiple applications especially in life science and other research fields. In this chapter and Chapter 5, we will consider applications of NV centers inside NDs.

As discussed in Chapter 1, NV center can respond to external signals through different mechanisms such as shift of resonant frequencies. In this chapter, we consider NV-based relaxometry, that is, the relaxation time of NV centers is correlated with environmental signals of interest. We start by introducing the longitudinal relaxation of NV centers and then proceed to the detection of magnetic molecules.

4.1 NV relaxometry: longitudinal relaxation

4.1.1 Overview

The longitudinal relaxation (T_1) process of NV centers is known to be sensitive to external magnetic noise, which has been used for noise spectroscopy and design various magnetic field sensors. In recent years, NV-based T_1 relxamoetry [62, 63] and similar methods have been developed to probe magnetic particles such as Gd complexes [64, 65, 66, 67, 68] in various environments, image electrical conductivity [69], detect localized chemical events [70] and free and coupled radicals [71, 72]. NV-based relaxometry has been a versatile technique with distinct advantages including nanoscale resolution, real-time measurements, and excellent sensitivity. In this section, we first give an introduction of the longitudinal relaxation process of NV centers in the presence of paramagnetic spins and then provide a simple model to calculate the relaxation time.

4.1.2 Model of T_1 relaxation

As mentioned in Chapter 1, a green laser (532 nm) can efficiently polarize the NV centers into the $|m_s = 0\rangle$ ground state. Due to the interaction with its environment, this spin polarized states will reach thermal equilibrium with a time scale T_1, known as longitudinal relaxation process (Fig. 4-1). Indeed, as we will show later, the presence of the external transverse magnetic noise will accelerate this process.

To provide a quantitative description of the T_1 relaxation, we analyze all the effects contributing to the longitudinal relaxation time. For shallow NV centers or NV centers in nanodiamond, these effects include the contributions from bulk lattice

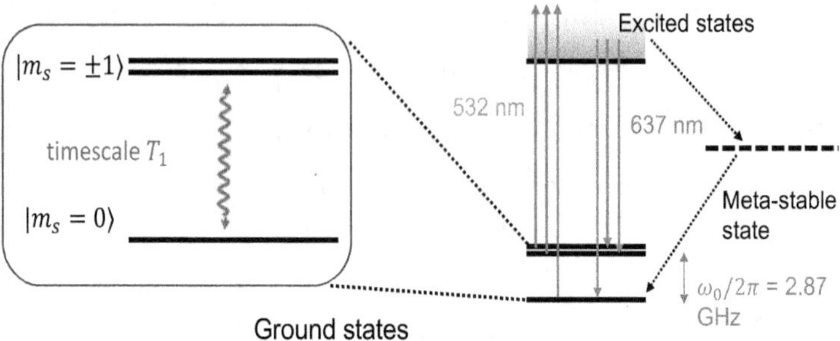

Figure 4-1: Right: Energy level diagram of an NV center showing the optical transitions in the absence of external static magnetic field. Green laser non-resonantly excites NV spin to the excited state, and the NV decays back emitting red fluorescence photons. The $|m_s = \pm 1\rangle$ states can also decay non-radiatively through the meta-stable singlet state and back to $|m_s = 0\rangle$ ground state during excitation (dashed lines), providing a mechanism for both optical initialization to $|m_s = 0\rangle$ and spin state-dependent optical readout. Left: spin polarized in one of the ground states will reach thermal equilibrium with a time scale T_1, known as longitudinal relaxation process. The presence of the external transverse magnetic noise will accelerate this process.

and paramagnetic impurities, and possibly other magnetic species that are present near the diamond surface. Without loss of generality, the T_1 time of a single NV electronic spin can be derived from the Fermi's golden rule [167, 4, 64]

$$\frac{1}{T_1} = \frac{1}{T_{1,bulk}} + \sum_k 3\gamma_k^2 B_{\perp,k}^2 \frac{\tau_{c,k}}{1 + \omega_0^2 \tau_{c,k}^2}, \tag{4.1}$$

where $T_{1,bulk}$ is the relaxation time of NV in bulk diamond at room temperature and we take it to be 1 ms hereafter, $\omega_0 = (2\pi)2.87$ GHz the NV energy level splitting between the $|m_s = 0\rangle$ and $|m_s = \pm 1\rangle$ states at zero external magnetic field. Here we consider the contributions of different species of magnetic particles. For each magnetic particle k, γ_k is the gyromagnetic ratio; $B_{\perp,k}$ is the rms transverse magnetic noise strength; and $\tau_{c,k} = 1/R_k$ is the noise correlation time, the inverse of the noise fluctuation rate.

In the following, we consider NV centers in small NDs with diameter d less than 100 nm. Then, the paramagnetic spins on the ND surface will significantly contribute to

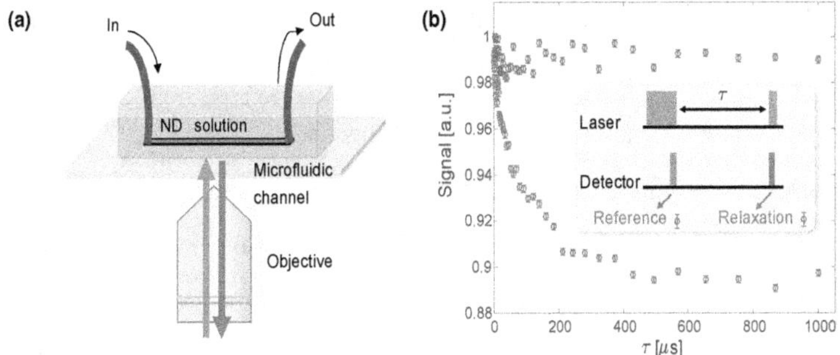

Figure 4-2: **(a)** Diagram of the experimental setup. ND solution with Gd molecules was loaded into a microfludic channel to prevent evaporation. We excited the ND sample with green laser and collected the red fluorescence with the same objective. **(b)** Typical T_1 measurement with reference and relaxation signal lines. The error bars for each data point are the standard deviations over repetitions. In the inset: T_1 relaxometry protocol.

the relaxation of NV centers inside due to the strong dipolar interactions. To evaluate the relaxation time, we need to evaluate both the transverse magnetic noise $B_{\perp,k}$ and the noise correlation time $\tau_{c,k}$ for different magnetic species k. Detailed calculations of these terms can be found in Appendix A.

4.2 Detection of magnetic molecules

The longitudinal relaxation of NV centers can be used to detect magnetic molecules such as gadolinium (Gd) complex, which are among the most commonly used contrast agents in magnetic resonance imaging (MRI). Gd molecules have 7 unpaired electrons, resulting to a spin number $\frac{7}{2}$. This yields a strong dipolar interaction between NV centers inside ND and free Gd molecules in solution, which then leads to the shortening of NV's T_1 time. In this section, we will present our results of using NDs hosting NV centers as sensors to detect the concentration of one type of Gd(III) complex, gadolinium 1,4,7,10- tetraazacyclododecane-N,N',N'',N'''-tetraacetate (Gd-DOTA) [167].

In this experiment, the NDs we used have an average size of around 25 nm and are terminated with carboxyl groups. We probe the relaxation time of ensemble of

NDs inside a microfluidic channel with a home-built confocal microscope at room temperature, without any applied external magnetic field. The details of setup and the sample characterization can be found in Appendix B. As shown in Fig. 4-2, the T_1 time is measured in an all-optical fashion and we fit the results with an exponential function to extract the timescale. Note that the relaxation time for individual NV centers might vary a lot due to inhomogeneities in the local environment of NDs. Indeed, one of the biggest challenges of using NV centers is reproducibility from one particle to the next [62, 63] and it's thus beneficial to use large ensembles of NV centers.

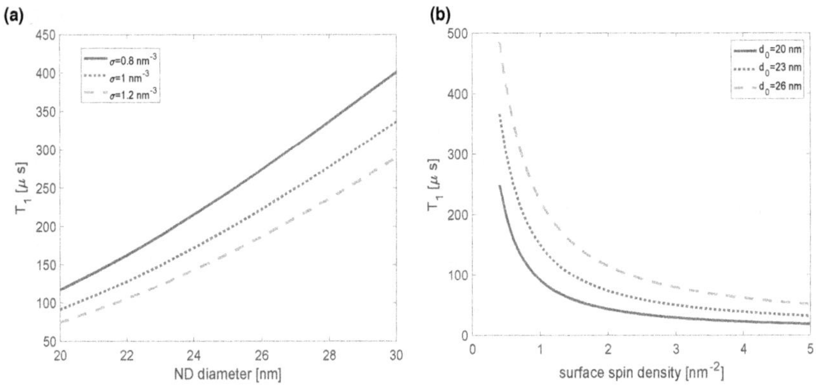

Figure 4-3: Estimation of NDs' surface spin density. Here we assume no Gd molecules were added. (a) T_1 as a function of ND diameter for fixed surface spin densities. (b) T_1 as a function of surface spin density for fixed ND diameters. From our measured relaxation time $T_1 \approx 133$ μs, we estimated $\sigma \approx 1$ nm^{-2} in our samples.

In the absence of Gd-DOTA, we measured a relaxation time of about 130 μs, which is significantly shorter than $T_{1,bulk}$ (typically a few ms, even in the presence of other bulk paramagnetic impurities). We attribute this difference to the bath of paramagnetic surface spins, for example, dangling bonds with unpaired electrons with spin 1/2 on the ND surface [168, 169]. The magnetic noise induced by these surface spins adds a new depolarization channel and yields a decrease in T_1 according to Eq. (4.1). Based on this model, we can further deduce the surface spin density to be about 1 nm^{-2}, as shown in the calculations in Fig. 4-3. T_1 relaxometry can then provide information in estimating surface spin densities of ND samples.

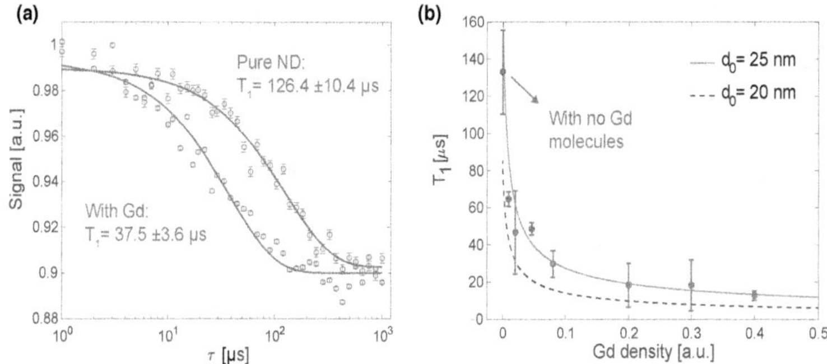

Figure 4-4: **(a)** T_1 quenching in the presence of Gd molecules. The experimental data shows the relaxation signal normalized by the reference, in the presence (red) and absence (blue) of Gd. We fit the decay to a single exponential (solid curves) and use the fitting error for the error bars of the data. **(b)** Relaxation time T_1 as a function of the density of Gd molecules (red). The bare ND relaxation time (blue) is around 130 μs. We measured the relaxation signal over several spatially separated spots and took the average of the fitted T_1. Error bars are the propagated fitting errors from the different spots. The experimental data matches the theoretical predictions with ND sizes of 20 nm (dashed lines) and 25 nm (solid lines), assuming higher Gd density than the prepared solution, due to aggregation of Gd molecules close to the edge of the microfluidic channel.

We next show the T_1 quenching effect induced by Gd molecules in the solution. As plotted in Fig. 4-4 (a), the presence of Gd molecules will induce a strong magnetic noise in addition to that arising from the surface spins, and significantly shorten the sensor's spin relaxation time. As expected, higher densities of Gd-DOTA molecules lead to stronger magnetic noise thus larger T_1 quenching ratio (Fig. 4-4 (b)). The experimental results can qualitatively match our theoretical predictions calculated from the model described in the previous section. We remark that contrary to many other NV-based biological sensors [65, 170, 171, 85, 172], microwave driving is not necessary here in principle, thus reducing the complexity and cost of the experimental setup. To characterize NV charge state fluctuations that might mask the real T_1 decay, fluorescence spectrum can be measured to extract the ratio of different charge states [70, 173, 6].

To this end, we have shown that the relaxation time of NV centers are sensitive to the presence of Gd magnetic molecules. A promising avenue to detect biological or chemical signals is to transduce them into magnetic noise – using, e.g., magnetic molecules, such as Gd complexes here, that NV centers can probe. In the following chapter, we will present how the strong interaction between NV centers and Gd complex can be leveraged for detection of various biological signals.

Chapter 5

Quantum sensing of biological and chemical signals

To detect biological and chemical processes, a significant challenge is that NV centers are usually not directly sensitive to the signals of interest, such as protein, glucose, gene sequences and ion species. These species usually don't have electronic spin labels that can strongly interact with NV centers inside diamond through dipolar coupling. A promising avenue to resolve this challenge is to transduce target signals into magnetic noise – using, e.g., magnetic molecules such as Gd complexes discussed above, or electrical noises – using, e.g., metal cations. This transduction usually relies on chemical engineering of diamond surface [87], such as attachment of stimulus-responsive polymers [70] and hydrogels [40]. It can greatly enlarge the application scope of NV-based sensing and leverage the advantage of NV centers including, for example, excellent sensitivity and high spatial resolution.

In this chapter, we first explore applications in detecting biological signals based on the T_1 relaxation process [167, 4]. The target signals are converted into magnetic noises of magnetic molecules that NV center is sensitive to. In the last part of this chapter, we present a design and demonstration of alkali cation sensor based on NV center's charge state fluctuations [174]. In this case, with the help of surface chemistry on nanodiamond, the ion presence is transduced into change of the local charge environment of NV centers.

5.1 Sensing of rotational Brownian motion

5.1.1 Sensor mechanism

We start by presenting a proof-of-principle demonstration of using NV center's T_1 relaxation to detect rotational Browninan motion of small molecules, followed by the measurement of Gd complexes discussed in Chapter 4. The intuition comes from the the model of T_1 relxation Eq. 4.1, where knowledge of the bath correlation time τ_c is needed in order to calculate the noise spectrum. For Gd molecules in solution that we detected, the total noise fluctuation rate $R_{Gd,tot} = 1/\tau_{c,Gd}$ is given by [175]:

$$R_{Gd,tot} = R_{Gd,dip} + R_{vib} + R_{trans} + R_{rot} \tag{5.1}$$

where $R_{Gd,dip}$ represents the dipolar interaction rates between Gd molecules, R_{vib} the intrinsic vibrational rate between the Gd ion's vibrational energy levels, and $R_{trans(rot)}$ the translational (rotational) Brownian motion rates in solution. Variations in the fluctuation rate $R_{Gd,tot}$ induce changes in the T_1 signal decay time of NV electronic spins via Eq. (4.1). It is then possible to detect, for example, the rotational Brownian motion (RBM) rates of the magnetic molecules and sense the local environment through T_1 relaxometry of NV sensors in an all-optical fashion.

Detecting variations in molecules' Brownian motions would yield information about particle size and local viscosity in biological sensing. For example, the random rotations of small molecules can respond to environmental changes and reveal local dynamics and biological functions. Conventional optical microscopy methods [176] are capable of measuring the rotational Brownian motions (RBM) of particles in the sub-micron scale or larger. However, it is very difficult to extend these techniques to the nanometer regime: the rotational motion of nanometer-size molecules is typically much smaller than the optical diffraction limit, therefore it cannot be directly captured via optical microscope; in addition, their motional rates are typically in the GHz range, which are beyond the detection rates of most optical techniques. To this end, the T_1 relaxometry of NV sensors inside small NDs would provide an all-optical method capable of capturing the fast rotations. The results presented below in this section is based on the paper Ref. [167].

As shown in Eq. 5.1, the fast RBM of Gd molecules R_{rot} contributes to high frequency fluctuations in the magnetic noise spectrum. Counter-intuitively, this added motion term actually often lengthens the relaxation time of NV centers. To understand how, we can consider a Lorentzian spectrum S as a function of the noise fluctuation rate $R_{Gd,tot}$, i.e., $S = \frac{\tau_{c,Gd}}{1+\omega_0^2 \tau_{c,Gd}^2} = \frac{R_{Gd,tot}}{R_{Gd,tot}^2 + \omega_0^2}$. When the fluctuation rate is larger than the resonant frequency of the spin sensors $R_{Gd,tot} > \omega_0$, a further increase in fluctuation rate results in a smaller noise spectrum intensity at the NV spin resonance frequency. This effect is similar to motional narrowing in NMR [177], and leads to a longer relaxation time of the spin sensors.

5.1.2 Experiment demonstration

To show the capability of detecting the RBMs of Gd-DOTA molecules with our sensors, we aim at distinguishing variations in the magnetic particle fluctuation rates due to changes in the solution viscosity. To change the viscosity of the solution, we dissolved a fixed density of Gd-DOTA molecules in solutions with varying concentrations of water and acetone. Varying the ratio between water and acetone in the binary solution changes the local viscosity felt by Gd-DOTA molecules, as shown in Fig. 5-1 (a). We perform T_1 relaxation measurements by loading the sample in a microfluidic channel to mimic biological environment, while preventing the binary solution from quickly evaporating (see Appendix B for details).

Note that the RBM rate is given by

$$R_{rot} = \frac{k_B T}{8\pi a^3 \eta f_r}, \qquad f_r = \left(\frac{6a_s}{a} + \frac{1 + \frac{3a_s}{a+2a_s}}{(1 + \frac{2a_s}{a})^3} \right)^{-1}. \tag{5.2}$$

Here f_r is the microviscosity factor that takes into account the discrete nature of solvent molecules [178, 179]. a is the radius of the Gd complex and a_s is the solution molecule radius and $a_s = 0.14$ nm for water (see Appendix A for details). As we can see, a lower viscosity corresponds to faster motions, with a RBM rate R_{rot} as large as

14 GHz in our experiment, as shown in Fig. 5-1 (a).

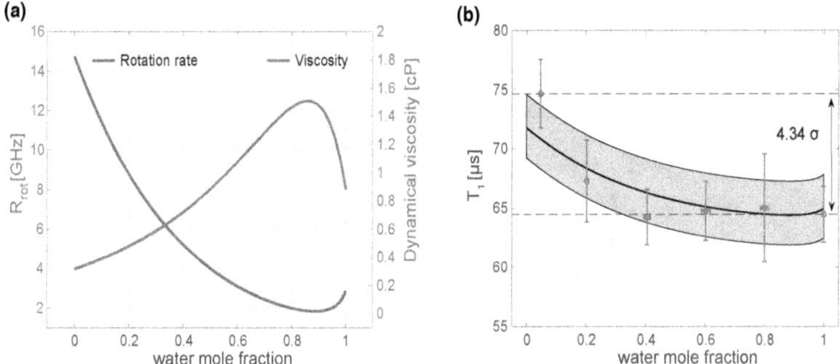

Figure 5-1:   (a) Predicted dynamical viscosity as well as motional rates R_{rot} as a function of water mole fraction in water-acetone binary solutions. (b) Experimental measurement of the relaxation time in different water-acetone binary solvents (the lowest water mole fraction we prepared was 0.046). The error bars are propagated errors from ensemble measurements over different confocal spots. The solid line corresponds to the theoretical estimation where the Gd molecule density is obtained by approximately matching the experimental result for (mean) T_1 in pure water with the theoretical prediction. The shaded area corresponds to 10% error for the Gd density.

From the experimental results, we observed a clear difference (4.34 σ) in the measured T_1 times for pure water with respect to (nearly) pure acetone (Fig. 5-1 (b)). This should be attributed to the different rotation rates R_{rot} that affect the magnetic noise spectrum, as mentioned before. Also, separate control test shows that NV center's T_1 time is not affected by the solution viscosity in the absence of Gd molecules. Since our measurements were performed with an ensemble of NDs, one concern was the inhomogeneous density distribution of Gd molecules when one changes solutions. To investigate this issue, we compared the distribution of relaxation times of spatially separated sensing spots for the water and nearly pure acetone cases. As one can see in Fig. 5-2 (a), the two distributions spread broadly, which might be a result of the spatial inhomogeneity of Gd molecule density or spatially varying local charge environments. Nevertheless, the two distributions are clearly distinct form each other, as shown in Fig. 5-2 (b), with Gaussian fittings. This further demonstrates that our relaxation measurements can distinguish different RBM rates influenced by local viscosities.

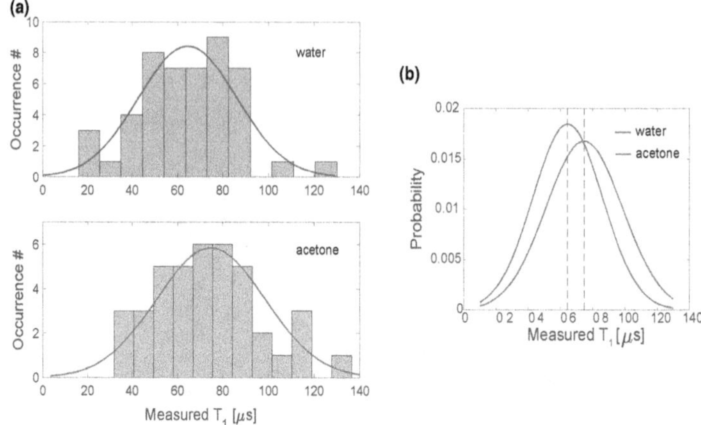

Figure 5-2: (a) Distribution of measured relaxation times over different confocal points for molecules in (top) pure water and (bottom) nearly pure acetone (with water mole fraction 0.046). (b) Gaussian fittings to the two distributions show two clearly distinct peak values.

To further quantify the performance of our proposed method, we can estimate the minimal detectable value of the total magnetic noise fluctuation rate $\delta R_{Gd}^{min}\sqrt{T}$ per unit time, for a single NV center located at the center of a single ND. We find (see Appendix A for calculation details)

$$\delta R_{Gd}^{min}\sqrt{T} \approx \frac{1}{C\sqrt{\mathcal{D}T_D}}\sqrt{\frac{2eR_{Gd,tot}}{3\gamma_e^2 B_{\perp,Gd}^2}}\frac{(R_{Gd,tot}^2 + \omega_0^2)^{3/2}}{|R_{Gd,tot}^2 - \omega_0^2|},\tag{5.3}$$

where T_D is the detection window, $\mathcal{D}$ the photon counting rate, and C the T_1 signal contrast. In our experiments we have a detection window $T_D = 500$ ns. We assume a signal contrast $C = 0.2$ and a photon counting rate $\mathcal{D} = 1 \times 10^5$ counts/s for a single NV. At optimized Gd density (corresponding to total Gd fluctuation rate $R_{Gd,tot} \approx 60.2$ GHz), we get a sensitivity of $\delta R_{Gd}^{min}=6.9$ (9.6) GHz for a single ND with diameter $d_0=20$ (25) nm in a $T = 10$ s data acquisition time, as presented in Fig. 5-3. Note that here the calculation only takes the photon shot-noise into consideration, as it is the main source of detection noise.

In the context of quantum sensing, we have shown that the T_1 relaxometry provides a versatile tool to detect the RBM rate changes in response to variations of

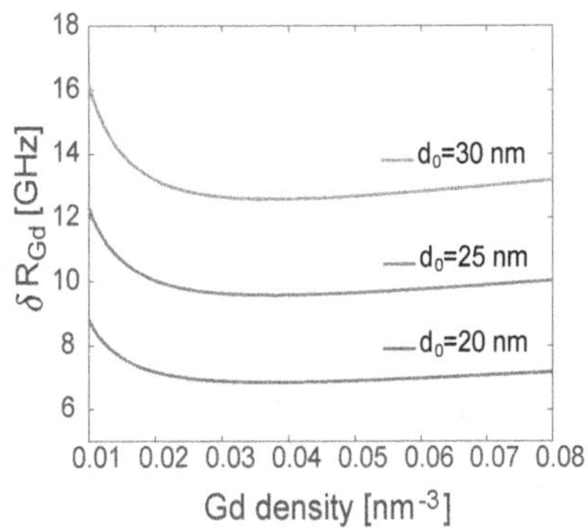

Figure 5-3: Sensitivity estimation as a function of Gd molecule density for a single NV sensor located at the center of a nanodiamond with diameter d_0 in a 10 s data acquisition time.

local environments. The technique is capable of detecting fast rotation (GHz-rate) of magnetic molecules with size down to sub-nanometer scale. The experiment is fully optical, requiring no microwave control or any external magnetic field. At the same time, we point out that the relation between sensor relaxation and rate variation can be employed in MRI techniques, where the fluctuation rates of contrast agents (usually magnetic molecules such as Gd(III) chelators) are modulated to increase their relaxivity and improve their performance [180].

5.2   Sensing of virus RNA

We have seen from previous discussion that NV centers are responsive to the presence of magnetic particles such as Gd complexes. Based on NV and magnetic molecules, various sensors have been developed to detect signals that NV center itself is not directly sensitive to. In Ref. [70], the system was used to image localized chemical events such as pH and redox potential changes based on cleavable polymers linking nanodiamonds and Gd molecules; by employing stimulus-responsive hydrogels as a spacing transducer in-between nanodiamond and magnetic Ni nanoparticles, in Ref. [40] a reversible temperature sensor was demonstrated. Indeed, these NV-Gd sensors that respond to external signal triggers via distance-dependent outputs, similar to fluorescence resonance energy transfer(FRET)-based sensors [181, 182], where energy is transferred from a donor fluorescent molecule to an acceptor molecule and the transfer efficiency has a strong dependence on the relative distance between donor and acceptor molecules. With the help of chemical engineering especially diamond surface chemistry [86, 87], biological and chemical signals of interest can be transduced into magnetic noise, thus leveraging the advantage of NV's excellent sensitivity in probing magnetic signals.

Inspired by this broad strategy, in this section, we present our design of a novel virus RNA sensor based on interaction between Gd molecules and NV centers inside NDs [4]. The sensor again uses an all-optical method and is based on T_1 relaxation measurement. Based on the discussions in Chapter 4.1, we evaluate the sensor's performance and find a detection limit as low as several hundreds of viral RNA copies in one second measurement time. We will show in the next sections that the sensor is expected to have false rates that are considerably lower than state-of-art polymerase chain reaction diagnosis method. Furthermore, the present diagnosis method is scalable, fast and low-cost, which can meet the requirements of accurate estimation of epidemic trajectories.

5.2.1 Background and detection mechanism

Introduction and motivation

The motivation of designing a NV-based quantum virus sensor origins from the Coronavirus disease 2019 (COVID-19) pandemic. The rapid spreading of the disease has shown the importance for rapid and cost-effective testing of emerging new viruses, as early diagnosis and population surveillance of the virus load are crucial to slow down and contain an outbreak. Developing accurate testing for novel viruses with low false negative rate (FNR) and false positive rate (FPR) is critical and enable taking appropriate preventive measures. While we take the detection of severe acute respiratory syndrome coronavirus 2 (SARS-CoV-2) as an example, the scheme introduced below can be generalized into diagnosis of other RNA viruses.

While the diagnostic testing field for COVID-19 and other virus disease is rapidly evolving and improving in quality [183, 184], in many cases it still cannot not match the demand. It's known that the SARS-CoV-2 has a single-positive strand RNA genome, which remains in the body only while the virus is still replicating, and can be detected by various virology methods, most preeminently by the reverse transcriptase quantitative polymerase chain reaction (RT-PCR). While RT-PCR has been the primary method of viral genome detection for SARS-CoV-2, it requires trained personnel, special equipment, and careful design of the primer and probe. More importantly, during PCR tests the samples need undergo an amplification process which can take several hours, and might degrade the diagnosis accuracy and the method can suffer from high FNR [185, 186]. The false negative results are particularly consequential since they might lead to infected persons not isolating and infecting others. To this end, a highly sensitive, accurate and rapid diagnosis method is on demand.

Sensor design

As discussed in the previous chapter, quantum sensors including NV centers in diamond have emerged as powerful tools to detect chemical and biological signals. We have shown that NV centers in nanodiamond are sensitive to external magnetic noise induced by magnetic molecules such as Gd complex. Along this line, we introduce a quantum sensor based on NV centers in NDs that is capable to detect virus RNA with high sensitivity.

Our proposed diagnosis technique is shown in Fig. 5-4. Firstly, viral RNA taken from an upper respiratory sample of patients is isolated and purified using fast spin-columns and this process is exactly the same as RT-PCR. Contrary to RT-PCR, reverse transcription is not be needed, as the sensing technique is based on direct detection of the virus RNA. Moreover, our sensor does not require nucleic acid amplification due to its high sensitivity to viral RNA, reducing the complexity, cost, and time of the test. The sample solution is directly ejected into microfluidic devices where the NV-based hybrid sensor is loaded beforehand.

We then move to the synthesis of the hybrid sensor. As we show in Fig. 5-4 (c), we first non-covalently coat NDs by cationic polymers [76, 187, 188, 189], such as polyethyleneimine (PEI), which can form reversible complexes with viral complementary DNA (c-DNA) sequences [190]. A stable c-DNA fragment of SARS-CoV-2 can be obtained by RT-PCR [191] or synthesis. The Gd^{3+} complexes such as Gd-DOTA can be incorporated into the sequence [3], forming hybrid c-DNA-DOTA-Gd molecules. Then, due to the molecular electrostatic interactions between PEI and c-DNA-DOTA-Gd^{3+}, the Gd^{3+} complexes will tend to lie on the ND surface, in close proximity to NV centers, thus efficiently increasing the magnetic noise strength felt by NV spins and quenching their T_1 time. We note that previous studies [1, 76, 192] have demonstrated that the ND-PEI-DNA hybrid nanomaterial is a stable and efficient gene or drug delivery systems, and survives both sonication and storage for several months. Therefore, the binding between c-DNA-DOTA-Gd^{3+} and ND surface should thus be stable against moderate temperature and mechanical fluctuations.

In the presence of viral RNA, the c-DNA-DOTA-Gd^{3+} pair will detach from the ND surface due to c-DNA and virus RNA hybridization. Note that the formation rate of c-DNA-RNA hybrids is considerably higher than the binding between c-DNA and polymer-functionalized ND [193, 194] and a detailed analysis can be found in Ref. [4]. Moreover, the c-DNA sequence on ND surface targets parts of the viral genome that are not considerably affected by naturally occurring viral mutations and are unique to SARS-CoV-2 [195] to ensure sustained efficacy and specificity. Now the newly-formed c-DNA-DOTA-Gd^{3+}/RNA compound will then diffuse freely in the solution, leading to an increased distance between Gd and ND. After appropriate centrifuging, the free Gd-containing compound can be separated from the NDs. The NV centers will then feel weaker magnetic noise and have a longer T_1 time, indicated by a larger fluorescence intensity at fixed wait time. By optically monitoring the change in relaxation time, we can identify the presence of virus RNA in the sample and even quantify the RNA number.

As mentioned above, while we consider the SARS-CoV-2 virus RNA in this work, the presented technology can be generalized to diagnosing other RNA virus such

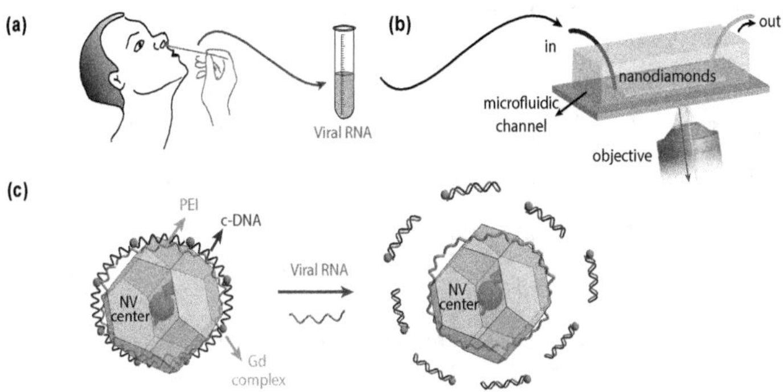

Figure 5-4: **(a)** A sample is collected from the upper respiratory tract, e.g., with a nasopharyngeal or throat swab, followed by nucleic acid extraction. **(b)** Test samples that might contain virus RNA are loaded into microfluidic channels containing functionalized NDs. The emitted red fluorescence signal resulting from green laser excitation of the NV is collected via a confocal microscope or on a CCD. **(c)** Mechanism of magnetic noise quenching. c-DNA is adsorbed on the surface of functionalized ND containing NV centers. The absorption is due to molecular interactions between cationic PEI polymer on ND surface and c-DNA. Other polymers such as poly-l-lysine could also be used to bind the c-DNA sequences. Gd^{3+} complex molecules that can induce strong magnetic noise are connected to the c-DNA structure. In the presence of virus RNA, the base-pair matching of c-DNA and RNA leads to detachment of c-DNA-DOTA-Gd^{3+} from the ND surface, resulting in weaker magnetic interaction between Gd^{3+} complex and NV centers inside the ND.

as HIV and MERS by using surface c-DNA that is specific to target virus. The technique can also be applied to detect single-strand DNA genome by replacing the c-DNA with an appropriate RNA sequence. Furthermore, while we consider NV centers in nanodiamonds, alternative quantum sensors or host materials might be adopted. For instance, sensors based on silicon-vacancy centers in silicon carbide might be developed.

As for scaling-up of our diagnosis method, we propose that single or ensemble NDs can be firstly incorporated in microfluidic devices with separated channels and samples that possibly contain viral RNA are then injected into the channels. Then the resulting fluorescence signals from all samples can be simultaneously collected by a charge-coupled device (CCD) camera, ensuring a high throughput diagnosis capacity. Furthermore, we remark that the sensor only involves synthesized NDs (ND synthesizing has been a mature technology and the cost is negligible for each measurement) and short sequences of genome, which greatly reduce the material cost.

We note that the proposed RNA quantum sensor can be potentially integrated with CRISPR technology to achieve even higher sensitivity. The relavant discussions can be found in the supporting material of Ref. [4].

5.2.2 Performance evaluation

Sensor based on single NV centers

To benchmark the performance of the proposed sensor, we then perform numerical simulations and analysis according to the T_1 relaxation model we introduced in Chapter 4.1. We first focus on the virus RNA sensor based on single NV centers. Note that our protocol sensitivity arises from the strong dependence of the T_1 time on the surface density of Gd molecules n, as shown in Fig 5-5. Thanks to the strong RNA-DNA hybridization compared to the DNA-PEI binding, we can assume that all the Gd molecules get detached in the presence of virus RNA and freely diffuse in the solution [4]. Then, a higher initial c-DNA-DOTA-Gd^{3+} density would yield a larger T_1 difference and a larger fractional change. We note that multiple Gd^{3+} complexes

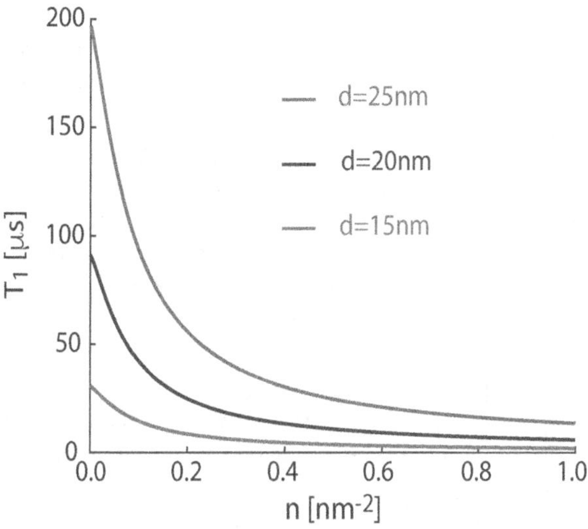

Figure 5-5: Relaxation time T_1 as a function of the surface density of Gd^{3+} complex molecules. The distance of Gd^{3+} complexes and ND surface is $l = 1$ nm. The plotted range of Gd density is a very conservative estimate based on previous studies of PEI ND coating [1], PEI-DNA binding [2] and DNA-Gd attachment [3].

can be incorporated into one c-DNA sequence [3] and the increased ratio between c-DNA and Gd^{3+} complexes will lead to an even larger change in T_1.

The above results show the potential for our proposed hybrid quantum sensor to detect the presence of the virus RNA. Indeed, the viral load can also be quantitatively estimated with high sensitivity and we benchmark its performance by calculating the quantum sensitivity $\delta n_{min}\sqrt{T}$ (see Appendix A for details) that is defined by the minimum number of detectable RNA copies in a total integration time T. In Fig 5-6 (a), we show the calculated sensitivity for varying parameters and an integration time $T = 1$s. As expected, a smaller distance between NV and Gd^{3+} complex (smaller ND diameter and Gd bound) would result in a better sensitivity.

We further show the distribution of quantum sensitivity (Fig 5-6 (b)) when taking into account variations in the NV sensor parameters that affect the sensitivity, including the randomness in ND diameter d, Gd^{3+} surface density n, as well as position of the NV centers with respect to the origin. As expected, these variations can give rise to broad distributions of the sensitivity, which predicts the probability that a

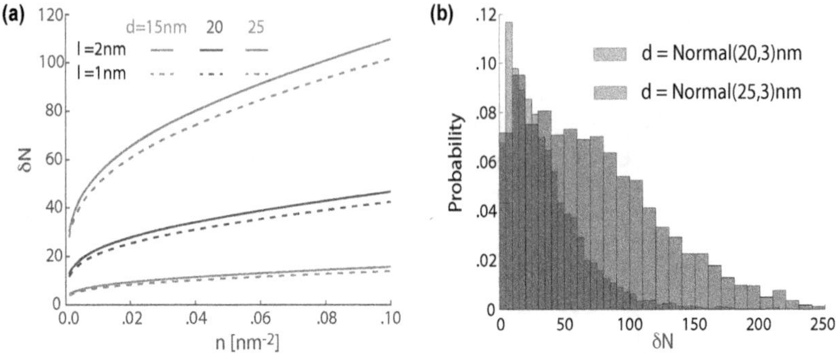

Figure 5-6: (a) Minimum detectable number of c-DNA-DOTA-Gd^{3+} molecules in T=1 s integration time (i.e., quantum sensitivity) for a NV center at the ND origin. Given the strength of the RNA-DNA hybridization compared to the DNA-PEI biding [4], we assume freely diffusing Gd molecules after detachment. At that point, their contribution to the T$_1$ can be safely neglected, as the distance between NV and Gd becomes significantly large. (b) Probability density distribution of quantum sensitivity (for optimal measurement time) considering the normal distribution of ND diameters, surface density of c-DNA-DOTA-Gd^{3+} as well as position of NV spin in the ND. The surface density n follows a Normal (0.1,0.02) nm^{-2} distribution. The distance between Gd^{3+} complex and ND surface is set to l=1.5 nm and the NV center position is randomly sampled in the ND. We fix the surface density of random paramagnetic centers at $\sigma = 1$ nm^{-2}, so the contributions from the ND bulk spins and surface spins to the relaxation process remain constant.

single hybrid sensor containing one NV center can detect a certain amount of RNA copies, without any pre-characterization. It's worthy to remark that when the sensor sufficiently interacts with the virus in sample, the minimal detectable number of RNA copies in 1-second integration time can be as low as 100. This is well below the viral load for positive clinical samples, which typically has $10^5 - 10^6$ RNA copies per throat or nasal swab [196] without nucleic acid amplification. Therefore, in this single-NV scenario, our sensor would reach a ultralow FNR ($<$1%) in contrast to the common RT-PCR diagnosis method [185]. We note that the ability to quantify the amount of viral load allows more accurately capturing the infectivity window and catching more contagion cases – even when the viral load is too small for other methods – thus playing an important role in accurate estimation of epidemic trajectories.

Sensor based on ensemble of NV centers

While measuring a single ND before and after introducing virus RNA can yield a considerably large T_1 difference, as we have demonstrated above, this protocol suffers from low photon counts and the challenge of addressing the same single ND. Moreover, for single NDs, the detectable number of RNA copies is upper-bounded by the number of c-DNA-DOTA-Gd^{3+} molecules on the ND surface, which is in turn limited by the surface area and surface density of c-DNA-DOTA-Gd^{3+}.

Therefore, in this section, we analyze a more practical scenario where the fluorescence signal from an ensemble of NDs is measured after a fixed dark time τ following the initialization laser pulse, which allows inferring the relaxation rate. In the simulations below, each NV center in the ensemble is characterized by different parameters, including its random position inside the ND, and (normal) distributions of the ND diameters and surface density of c-DNA-DOTA-Gd^{3+} around their nominal values.

As shown in Fig. 5-7, we plot the distribution of normalized photoluminescence counts at a fixed waiting time τ for a large number of NDs. The result is classified as negative for the presence of the virus when the observed photon count is below a certain chosen threshold (since it indicates a smaller relaxation time T_1) and vice versa. We set the threshold by maximizing the *balanced accuracy* = 1- (FNR + FPR)/2 from the calculated PL distributions, which describes the average of sensitivity (1-FNR) and specificity (1-FPR). When a misclassification occurs (e.g., the photon count is low even if virus RNA was present.), one gets either a false negative or a false positive case. It's then clear to see in Fig. 5-7 (a) that due to the variation in ND diameters and other parameters, the two count distributions arising from single NVs have a considerable large overlap, leading to FNR(FPR) = 0.14(0.239). This issue can be resolved by measuring an ensemble of NDs simultaneously. It can be seen from Fig. 5-7 (b) that as few as 10 NDs can reach an accuracy > 99.6 % with FNR(FPR) < 0.1(0.9)% demonstrating a much lower potential FNR of our quantum sensor compared with the common RT-PCR method.

Then, to further show the performance of the quantum sensor in ensemble mea-

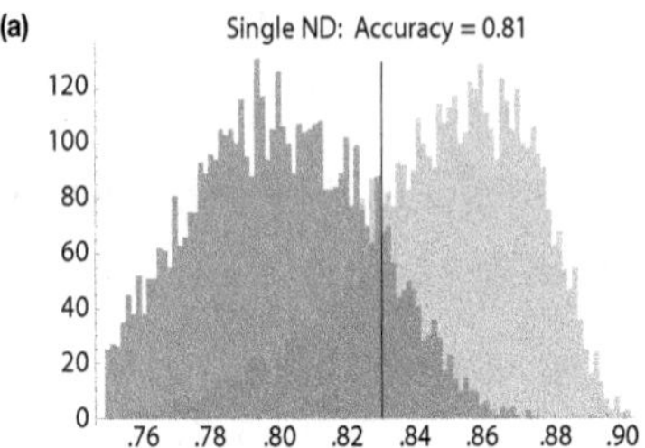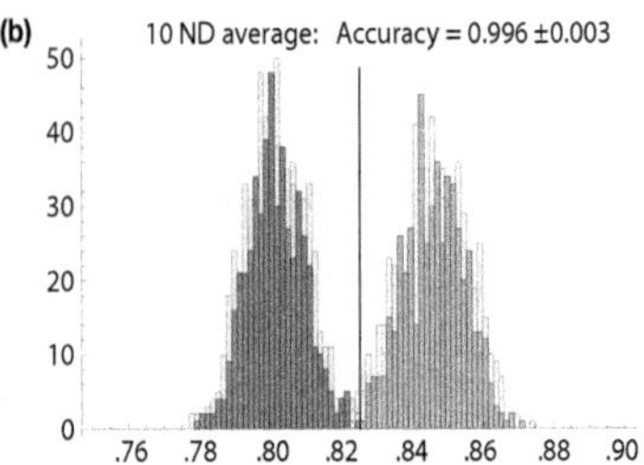

Figure 5-7: **(a)** Histogram of measured PL from single NDs at a fixed dark time $\tau = 200$ μs for average ND diameter $\bar{d} = 25$ nm. The NV position is random in a sphere of 20% of the ND radius. The red (green) distribution corresponds to the case where viral RNA is absent (present). **(b)** The NDs in (a) are grouped into random ensembles of 10 NDs and averaged over. The histogram with deep (light) colors shows the PL without (with) photon-shot noise. In both figures (a-b), the black lines indicate the optimal threshold that gives the maximum accuracy.

surements, in Fig. 5-8 we present the FNR (FPR) as a function of the number of SARS-CoV-2 RNA copies that can be detected by a group of NDs. When a larger number of NDs are measured simultaneously, the PL distributions are more well-resolved, leading to lower FNR (FPR) as expected. The effects of photon-shot noise on the photon count distribution overlap are shown in two extreme cases: the photon shot noise can indeed either add or subtract to the (average) photoluminescence signal. Correspondingly, the noise moves the signal distributions close to each other (worst case, shown in solid lines) or separate them further away (best case, shown in dashed lines). Even when considering photon shot noise, the FNR(FPR) achievable is still outstanding (only the largest diameter ND are considerably affected by the noise, since their average PL distributions are narrower).

To further improve the performance of the sensor, optimization of ND parameters would be beneficial. Random positions of NV centers inside ND lead to wide distribution of signals and NV centers near ND surface would also significantly suffer from random surface charge noise, potentially leading to deleterious charge dynamics of the NV centers. To resolve this issue, one may perform surface coating [197, 198] to produce NV centers that suffer less from surface noise and a pre-characterization of NVs' local charge environment would also be beneficial.

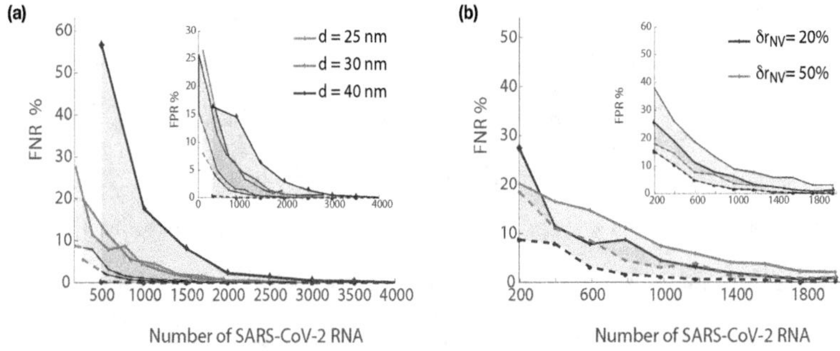

Figure 5-8: **(a)** FNR (inset: FPR) as a function of number of SARS-CoV-2 RNA copies associated with ensembles of NDs with different diameters. The solid (dashed) curves show the worst (optimal) case where the photon-shot noise increases (decreases) the PL distribution overlap. The NV position is random in a sphere of 20% of the ND radius. **(b)** Same as (a), but here we compare the effect of reducing the uncertainty in the NV position from 50% to 20% of the ND radius (ND average diameter 25 nm). In plotting the distributions in (c-d), we consider 5000 NDs with one NV each, and 0.1 nm^{-2} average surface c-DNA-DOTA-Gd^{3+} density. The ND diameter has a normal distribution with variance 3 nm and the average distance between ND surface and Gd molecules is 1.5 nm, with 0.2 nm variance.

5.3 Charge state sensing of alkali cations

In the previous sections, we have shown that NV centers in NDs are capable of probing external magnetic noise. Indeed, NV centers are also highly sensitive to their surrounding charge environments [6, 73, 74, 77]. Depending on the local charge distributions and external electrical manipulations, the NV state could have dynamical transitions between the widely-studied negatively charged state NV$^-$ and a neutral state NV^0 [5]. One extra electron of NV$^-$ can be ejected into the conduction band, resulting in NV^0 state; reversely, the NV^0 state can acquire one electron from valence band and becomes NV$^-$. The two charge states can be distinguished from their PL spectrum. For example, NV$^-$ has ZPL wavelength at 637 nm while it's 575 nm for NV^0. They also have very different phonon side-bands [6, 5]. Therefore, the fraction of different NV charge states can be extracted from a simple PL measurement in an all-optical fashion.

The charge state of NV centers can then serve as a meter to yield information about their environment. Indeed, previous studies have used NV charge state to probe functionalized groups [74, 75], DNA molecules [76, 77] on diamond surface as well as applied voltage on diamond [73, 42]. Therefore, the charge state of NV centers provides a new tool for characterizing and sensing chemical species or processes that can modify charge environment of NV defects inside diamond.

Along this line, in this section, we will present our work on designing and demonstrating an alkali ion sensor based on the modification of NV centers' charge states [174]. By covalent grafting crown ether structures on the surface of NDs, we build sensors that are capable of detecting specific alkali cations. As a proof-of-principle example, we use 15-crown-5 to detect sodium cations. The formation of 15-crown-5-Na^+ complexes layer on nanodiamond surface will sufficiently modify the charge environment of the NV centers inside and thus alter the charge states of NV defects. We then perform photoluminescence measurement on the emission spectrum of the NV defects in NDs under continuous green laser illumination, from which the fraction of different NV charge states are extracted.

5.3.1 Motivation and detection mechanism

Alkali ions such as sodium and potassium are very important metal ions in biological systems and their concentration is tightly regulated and can vary in different bodily fluids. For example, sodium and potassium ions are responsible for maintaining fluid and electrolyte balance and typical Na^+ (K^+) concentration in human blood is 135-150 mM (below 5 mM), while it's less than 30 mM (about 150 mM) in intracellular fluid [199, 200]. Their fluctuations are usually problematic and can trigger and indicate various physiological disorders and diseases, including cardiovascular disease and hypertension [201, 202]. Measurement of their concentrations would thus be of great interest both in understanding the functions of the ions in cellular physiology and for clinical diagnosis.

Clinical laboratories typically use ion-selective electrodes [203] or flame photometry [204] to perform measurements, but a sample volume as large as several mL would be required. To resolve this issue, and to achieve a high spatial and temporal resolution, small molecule-based fluorescence sensors would be favorable. In recent years, developing small molecular or nanoparticle probes for sodium or potassium ions has been an attractive research area [199, 200, 205]. Unlike instrumentation methods in clinical labs, they could not only provide a good spatio-temporal resolution but also potentially cross the cell membrane and be used for intracellular measurements.

One of the key challenge for designing such ion sensors would be to distinguish the target from other common metal ions. As an example, a sodium sensor should be able to tell the difference between sodium (Na^+) and other high concentration ions in biological system, such as potassium (K^+), magnesium (Mg^{2+}) and calcium (Ca^{2+}) ions. Along this line, the crown ether family has been widely used in designing alkali metal sensors thanks to its capability of binding alkali ions selectively [206]. They are cyclic chemical compounds that consist of a ring containing several ether groups, which can bound specific cations. For example, 18-crown-6 is known to have a high affinity for potassium cation while 15-crown-5 for sodium cation and 12-crown-4 for lithium cation. Using crown ether and together with a fluorescent moiety, highly sensitive metal ion sensor can be then built [206, 207].

We now present our work on detecting sodium cations based on monitoring the charge state of NV centers inside crown ether-functionalized NDs. We start by introducing the basic detection mechanism of the sensor. As shown in Fig. 5-9 (a), fluorescent ND with initial carboxyl group terminations is firstly coated with crown ether via covalent bond. This yields a neutral surface layer in contrast to the previous negative layer due to the carboxyl groups. Due to the chelate effect and macrocyclic effect, depending on its cavity size, crown ether exhibits strong affinities for specific alkali ions [206, 207]. Upon introducing these ions, stable complexes will form which then in turn leads to positive charges on ND surface. For example, the 15-crown-5 structure we studied in this work can strongly bond Na^+ and form complexes, while its interaction and affinity with other cations being relatively weak. While we focus on 15-crown-5 and sodium cation in this work, the scheme is general and can be extended to probe other alkali or alkaline earth metal ions using different types of crown ethers, as shown in Fig. 5-9 (b).

Then, the accumulated charges on the ND surface can effectively result to bending

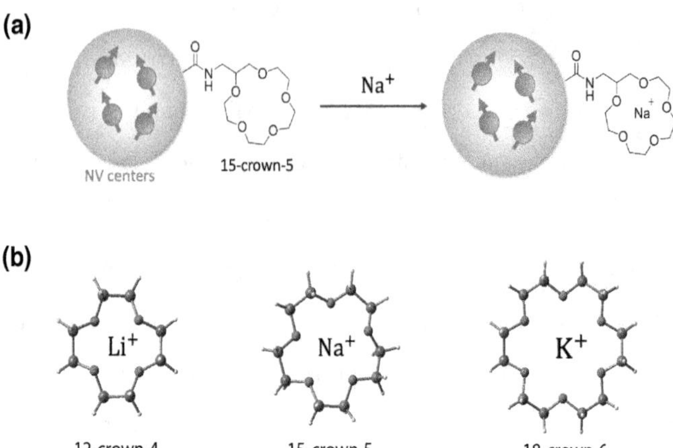

Figure 5-9: **(a)** Schematics of the ion sensor. NV center-containing NDs are firstly coated with crown ether. Upon their introduction, sodium cations will be trapped in the 15-crown-5 compounds. **(b)** Generalization into other type of crown ethers.

of the valence band and conduction band in the diamond lattice [73, 74, 77] (Fig. 5-10). Similar as hydrogen-terminated diamond [208, 74], a positive charge layer on the surface can lead to upward bending of both bands and the bending of the $NV^{-/0}$ level, which indicates the energy at which the NV defect loses or takes up one electron. In other words, this transition level indeed corresponds to the transition from the neutral to the negatively charged NV states. It's predicted to be in the band gap and 2.8 eV above the valence band maximum [5]. When its relative position with respect to Fermi level is lower (higher), the defect tends to absorb (lose) one electron.

Due to the positive charge layer on surface, when one NV is close to the surface, this transition level is shifted above the Fermi level and then the NV defect is ionized from NV^- to NV^0. At larger depths, instead, the band bending effect is weak and the Fermi level is above the $NV^{-/0}$ level due to the abundance of electron donors in diamond lattice, so the dominant charge state would be NV^-. We note that the NV^0 can further lose an electron and become a non-fluorescent and spinless state NV^+ [5], which cannot be detected here. In short, the formation of 15-crown-5-Na^+ complex leads to change of NV^- fraction, which can then be readout by performing

PL spectrum analysis.

5.3.2 Simulations on band-bending effect

We first perform numerical simulations to quantatitively study the band-bending effect using the nextnano software [209], a tool for simulation of electronic semiconductor nanodevices. Specifically, we rely on three-dimensional simulations on a ND of spherical shape with a diameter $d = 40$ nm (the average diameter of NDs used in our experiments). In the simulation, the Poisson equation is discretized on a grid with grid size 0.5 nm using the finite differences method and solved iteratively in a self-consistent manner. The Poisson equation is coupled with the single-band effective mass Schrödinger equation via the charge density, as described by the wave functions in the diamond [73, 74]. Negative charge (electron) donors in the lattice mainly include the substitutional nitrogen atoms (ionization energy 1.7 eV) and here we take their concentration to be 100 ppm (as expected from the ND used in experiments) and assume uniform distributions as a function of depth below the diamond surface. We take the nitrogen-to-NV conversion ratio to be 1%, corresponding to a total NV defect concentration of 1 ppm. More details about the simulation can be found in Ref. [174].

In Fig. 5-10 (a-c) we present the simulated band bending effect due to sodium cations on the surface of ND with a diameter being 40 nm. It's found that when the surface ion density is (0.1) 1 nm^{-2}, at a depth of ($<$1) 12.5 nm the dominant species of NV defect switches from neutrally to negatively charged states. We note that when the ion density is high, close to the surface, the valence band is shifted above the Fermi level, indicating p-type surface conductivity due to the form of a two-dimensional hole gas.

After getting the relative position of the charge transition level $E_{-/0}$ with respect to the Fermi level E_F, we use the Fermi-Dirac distribution to extract the average fraction of NV$^-$ charge state at depth x:

$$\langle n_-(x) \rangle = \frac{1}{1 + e^{(E_{-/0}-E_F)/k_BT}}$$

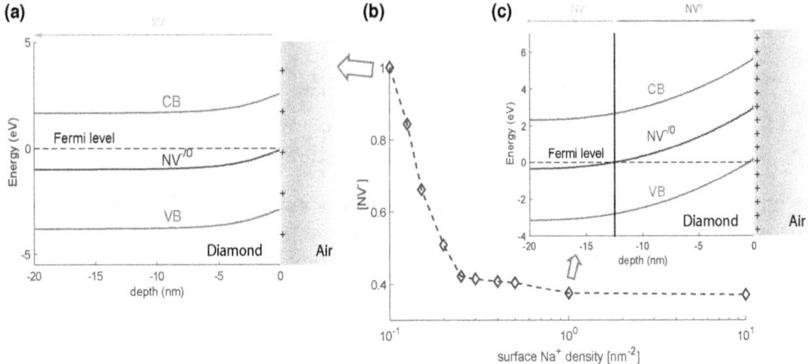

Figure 5-10: Simulations of band-bending effect. Positive charges on the diamond surface lead to upward bending of the conduction band (CB) and valence band (VB) as well as the $NV^{-/0}$ transition level (in the band gap and 2.8 eV above the valence band maximum [5]). The dashed line indicates the Fermi level. NV^0 is the dominant NV charge state when the $NV^{-/0}$ transition level is above the Fermi level, while NV^- dominates in the opposite case. The surface cation density is 0.1 nm^{-2} in (a) and 1 nm^{-2} in (c). Plot (b) shows the total NV^- fraction as a function of surface ion density.

where k_B is the Boltzmann constant and T = 300 K is the temperature. The total fraction of NV^- in the whole ND lattice with diameter d can then be calculated by performing the integral

$$[NV^-] = \int_0^{d/2} \langle n_-(x) \rangle g(x) dx$$

where $g(x)$ is the normalized NV defect density profile in the diamond lattice, i.e., $\int_0^{d/2} g(x)dx = 1$. In our simulations, we assume uniform distributions of NV defects.

Then one can extract the total fraction of NV^- as a function of surface cation density (here we use sodium ion as an example). As expected, in Fig. 5-10 (b) we observe that a larger density of cations on ND surface will serve as electron acceptor, thus leading to a smaller NV^- fraction in the diamond lattice.

5.3.3   Experiment results

Chemical synthesis and characterization

To demonstrate the above mechanism, we coat NDs with 15-crown-5 via EDC coupling and use the sensor to detect sodium cations. Following reported procedures [210, 211], NDs were first treated with a mixture of acids to remove the surface graphite contaminants and metallic impurities, generating carboxyl groups for the subsequent functionalization by EDC coupling. In particular, NDs were dispersed in a 3:1 mixture of concentrated sulfuric acid and nitric acid and stirred overnight at room temperature. After neutralization with aqueous NaOH (1 M), the resulting NDs was cleaned using several centrifugation, washing, and redispersion cycles with Milli-Q water.

EDC coupling with amino crown ether was performed by adding an excess of 1-(3-dimethylaminopropyl)-3-ethylcarbodiimide hydrochloride and 2-aminomethyl-15-crown-5 to the NDs dispersion and the reaction mixture was stirred overnight at room temperature, followed by several centrifugation, washing, and redispersion cycles with Milli-Q water. The resulting 15-crown-5 coated NDs (labelled as NDCE5 hereafter) were characterized by a Bruker Alpha II FTIR spectrometer with a Diamond Crystal ATR (attenuated total reflectance) accessory and a K-alpha+ X-ray Photoelectron Spectrometer system (Thermo Scientific) using a Al Kα radiation source (Fig. 5-11).

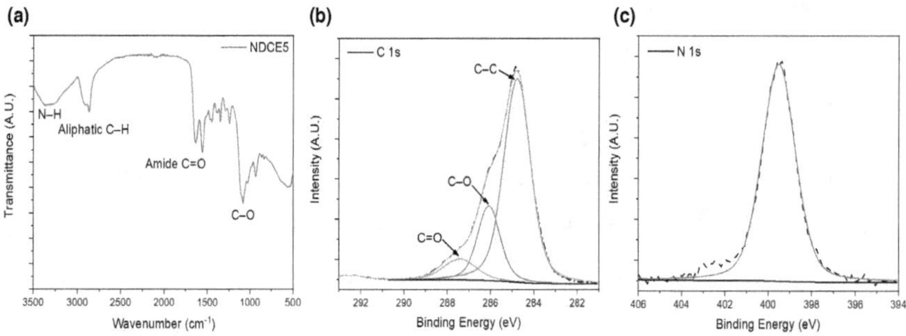

Figure 5-11:   Characterization of 15-crown-5 coated NDs (NDCE5) sample. **(a).** FT-IR spectrum of NDCE5. **(b)-(c).** XPS C 1s and N 1s spectra of NDCE5.

PL measurement

In order to measure the change of NV$^-$ fraction induced by surface charge, we perform PL spectrum measurement on ND ensembles under 532 nm laser illumination using a confocal Raman microscope (Renishaw inVia microscope with a 1024 × 256 pixel CCD camera) at room temperature. The samples were put on a silicon wafer to avoid unwanted Raman peaks.

It's known that NV$^-$ and NV^0 have distinct PL spectrum [6, 212]. For example, the zero-phonon-line (ZPL) for NV$^-$ is 637 nm while it's 575 nm for NV^0 as mentioned. To approximately extract the fraction of NV charge state we first peak-normalize the measured curves and then perform linear fitting of the acquired spectrum I(λ) from 560 nm to approximately 750 nm based on the single NV$^-$ and NV^0 reference spectra [6]. That is,

$$I(\lambda) = a\{[NV^-]I_- + (1 - [NV^-])I_0\}, \tag{5.4}$$

where $I_{-(0)}$ denotes the peak-normalized spectrum for a NV$^{-(0)}$ and a is a normalization factor.

In Fig. 5-12 (a) we give an example of a typical spectrum for carboxyl-terminated ND measured under continuous 532 nm laser illumination. The fitting here yields the fraction of NV$^-$ being [NV$^-$] = 0.754, which is comparable to the value observed in bulk diamond [213, 214].

Control test

As a control test, we first demonstrate that the charge state of NV centers in the carboxyl-terminated ND (labelled as ND in the plots) is not sensitive to sodium cations. Firstly, the ion itself shows no fluorescence under 532 nm laser. Then, as plotted in Fig. 5-13 (a), the distribution of [NV$^-$] shows little dependence on the concentration of sodium ions. The laser power here is fixed to be 0.5 mW. We further present typical illumination power dependence curves of [NV$^-$] for these NDs in the absence of external ions (Fig. 5-13 (b)). The green excitation dynamically modulates

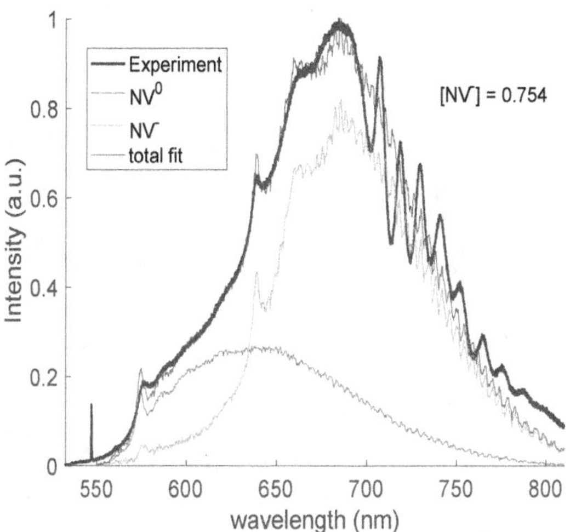

Figure 5-12: Typical PL spectrum for unfunctionalized NDs (carboxyl-terminated) acquired under 532 nm laser illumination. We perform linear fitting of the spectrum to extract the fraction of NV^- charge state. The peak-normalized $NV^{-(0)}$ spectrum is from reference [6]. The Raman peak at around 547 nm is attributed to the silicon wafer over which we deposit our samples. Oscillations of PL intensity after 700 nm are due to the etaloning effect on the CCD camera.

the NV charge state between neutral and negative. Due to NV's successive absorption of two photons with wavelength less than 637 nm (corresponds to 1.946 eV), an extra electron of NV^- can be ejected into the conduction band. The first photon pumps the NV^- from ground state to excited state while the second photon takes the electron to conduction band. In the opposite case, a neutrally charged NV defect can take up one electron from the diamond valence band band via absorption of a photon with energy larger than 2.156 eV (wavelength 575 nm). Under 532 nm illumination here, both the electron-NV recombination and the NV^- ionization process can happen, yielding an equilibrium NV^- fraction around 0.7. Again, this is similar to NV centers in bulk diamond [213, 214]. To further study the power dependence of the recombination or ionization process separately, another laser with longer wavelength (for example, 632 nm) is required to induce unidirectional charge conversion process while suppressing the reverse one.

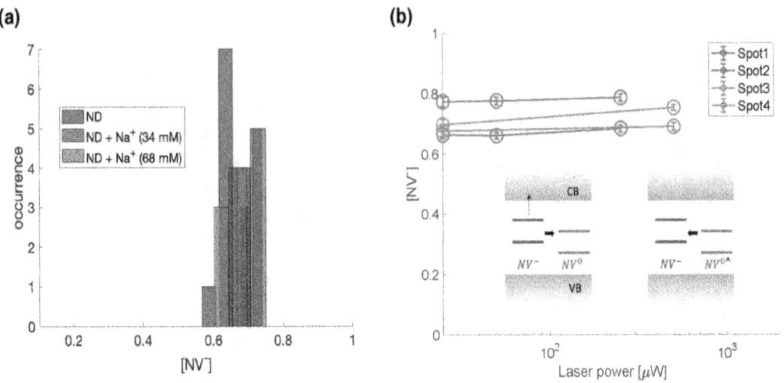

Figure 5-13:   (a) Distribution of NV$^-$ fraction for carboxyl-terminated NDs in the absence and presence of sodium ions. The laser power is fixed to be 0.5 mW. (b) Typical laser power dependence of NV$^-$ fraction of carboxyl-terminated NDs for four different spots. The errorbars are the fitting errors (5 percent). Inset shows the recombination and ionization processes of the two NV charge states.

Comparison of samples

We then switch to the crown-ether-functionalized ND sample (labelled as NDCE5) and measure its fluorescence spectrum under the same 532 nm laser illumination. The measurement was performed from low laser power to high power chronologically. It's clear shown in Fig. 5-14 that now the emission spectrum for NDCE5 is power-dependent. When one increases the laser power, the spectrum shifts rightward, corresponding to an increasing NV$^-$ fraction. We note that overall the fraction of negative charge state for NDCE5 is smaller than unfunctionalized ND even when the laser power is high. This can be attributed to the fact that NDCE5 is supposed to have neutral surface, while the carboxyl-terminated NDs tend to have negatively charged surface.

For a comparison, we then show the the power-dependence of [NV$^-$] for different samples in Fig. 5-15. While for unfunctionalized NDs we observe no dependence on illumination power, for all crown-ether-coated samples studied in this work, we see that when the power is low the NV$^-$ fraction is approximately zero and it then keeps increasing as a function of illumination power. Similar phenomenon has been observed in near-surface shallow NV centers [214] and this is in contrast to NV centers in some

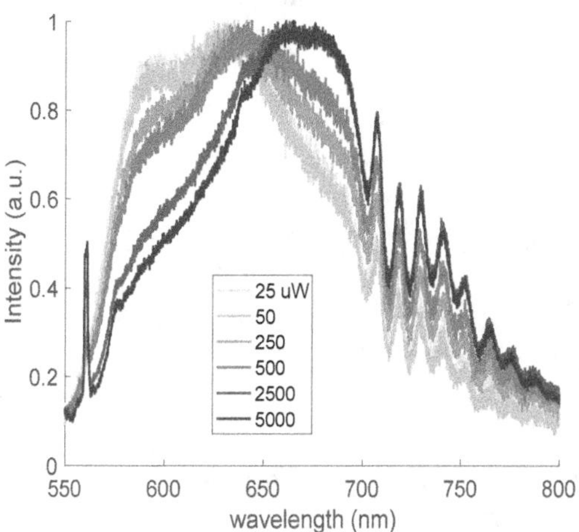

Figure 5-14: Typical PL spectrum of NDCE5 under different illumination powers.

previously-studied bulk samples [212], where the NV^--to-NV^0 ratio decreases as a function of the laser intensity. The power dependence observed here emerges from the interplay between the ionization and recombination rates and the charge transfer rate from NV^- to neighboring traps [214, 215]. These trap states are likely originated from neighboring defects such as vacancy complexes.

While the NDCE5 curve typically shows a saturated fraction around 0.5, after adding sodium cations, due to the positive charge on ND surface and thus the band bending effect, the NV^- fraction is considerably lower than 0.5 (usually below 0.4) even at high laser power, as it can be seen from Fig. 5-15 (a). We further show the data for NDCE5 in the presence of potassium ions. In contrast to sodium ions, 15-crown-5 has low affinity for potassium cation and the sample shows a high $[NV^-]$ at large laser power.

Finally, in Fig. 5-15 (b) we summarize the results for various samples we evaluated. Here we plot the absolute change of $[NV^-]$ as a function of the maximal $[NV^-]$ the sample can achieve at high illumination power. We note that in the presence of sodium ions, the maximal fraction is considerably lower than other samples, including NDCE5 and unfunctionalized NDs. Again, this is due to the fact that 15-crown-5 can form complexes with sodium cations, leading to the band bending effect and thus to

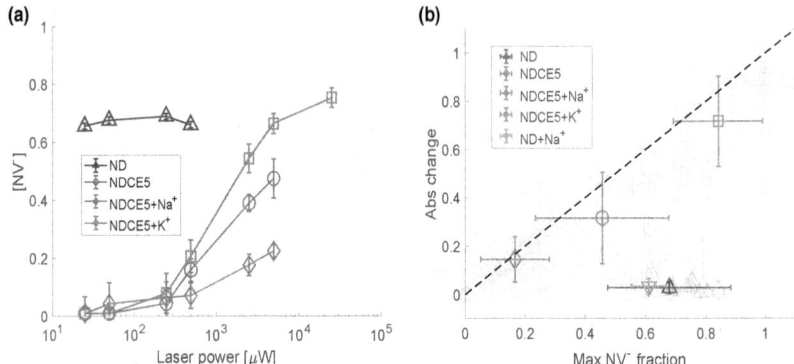

Figure 5-15: (a). Typical laser power dependence of NV^- fraction for various samples. Measurements are performed in order of increasing power. (b). Comparison of different samples using the coordinates of maximal reachable NV^- fraction and absolute change of NV^- fraction when one varies illumination power. The lighter data points show the results for individual spots, while the opaque data is the average results of different samples. The dashed line is when the change of NV^- fraction equals the maximal, i.e., when starting from $[NV^-]=0$ at low power.

a lower NV^- fraction even when the laser illumination power is high. Surprisingly, we find that in the presence of K^+, at high power most of the NV defects can be converted to NV^-. The reason for the phenomenon is not clear from the presented data and further study is needed to investigate the interaction between 15-crown-5 and potassium cations [216].

We further notice that a memory effect for the NDCE5 samples (both with and without sodium) studied above is observed. That is, after finishing the spectrum measurement at high laser power, we set the power to low values and re-evaluate the properties of spectrum. Fig. 5-16 shows that the NV^- fraction remains at a high value (above 0.5) even when the laser power is set as low as 50 μW. The memory effect persists for at least hours. While a detailed study on the charge dynamics and electron transfer process is needed, we suspect that high laser power preferentially transfers an excess electron from local traps to the NV defect [214].

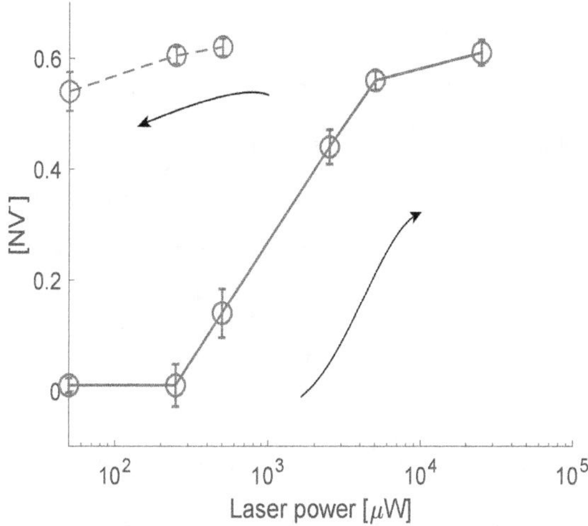

Figure 5-16: Typical memory curve of NDCE5 sample (here we show a case without sodium or potassium cations). The solid curve were first measured in power ascending order and then we went back to low powers (dashed line, in power descending order).

Remark and summary

We note that the due to the lack of ability in repeatably addressing the same ND in our experiments, we have to perform measurements over many spatial spots to average out the inhomogeneities among NDs and get the distributions of different samples to extract their difference. Similar as the virus RNA sensor we discussed in the previous section, the inhomogeneities can be induced by various factors, including inefficient surface coating of crown ether, spatial density profile of NV centers and nitrogen atoms as well as distinct local charge environments. This distribution measurement process is time costly and can greatly reduce the sensitivity of the sensor. To overcome the problem, we propose that NV centers with pre-characterized local charge environments and surface charge densities might be used to reduce the overhead in measuring many spatial spots.

We remark that while we use an all-optical approach to readout the charge state information of NV defect via the PL spectrum measurement, one might extract the same information from the signal contrast of optical-detected magnetic resonance

(ODMR), given the fact that NV^0 and NV^- have different ground state configurations. Larger NV^- fraction will yield a better signal contrast at the resonance frequency (2.87 GHz in the absence of external magnetic fields). For ODMR experiment, microwave pulses would be required. With the help of microwave, the surface charge layer might also be probed by monitoring the transverse relaxation time of the NV^- charge state, as it can induce fluctuating electrical field that can interact with the NV centers inside ND [43, 44].

In summary, in this section we present our work on designing alkali cation sensor based on NV centers in NDs with crown ether surface terminations. The charge state of NV centers inside shows a strong dependence on the surface charge profile and can be detected by measuring the PL spectrum of NV centers. As a proof-of-principle demonstration, we synthesize 15-crown-5-coated NDs to detect sodium cations. The proposed work here might open new opportunities for monitoring ion concentrations in biological systems and for cellular physiology.

Chapter 6

Concluding remarks

In this thesis we present our efforts in finding intriguing applications based on nitrogen-vacancy centers in diamond, a spin defect system that is subject to high controllability even at room temperature. We start by giving an introduction of the NV system and an overview of recent developments in NV quantum devices. Based on it, we explore two interesting research directions, quantum geometry measurement and quantum sensing.

On the fundamental physics side, we show that one can engineer a four-dimensional Weyl-like Hamiltonian via microwave control on the triplet ground state of a single NV center. By measuring the quantum geometric tensor of the ground state of the Hamiltonian via weak periodic modulation method, we successfully reveal the existence of a tensor monopole at the origin of the four-dimensional parameter space. The tensor monopole and the associated tensor gauge field is characterized by the $\mathcal{DD}$ invariant, a generalization of the celebrated Chern numbers in higher-form gauge field. The synthetic Hamiltonian model could serve as a playground for exploring unconventional quasiparticles beyond Dirac and Weyl fermions in high dimensional space [139, 140]. Moreover, from the viewpoint of high energy physics and exotic gauge structures, our model provides a testbed for novel non-Abelian gauge theories, especially in the case that the system has degenerate nodal rings.

While we only use the NV center in this experiment, by introducing strongly or weakly interacting spins such as ^{13}C and nitrogen atoms near the central NV spin, our work would pave the way for studying geometric properties in quantum few-body or many-body systems. Possible directions along this line include exploring the interaction between tensor monopole with other exotic gauge structures. Moreover, our experiment opens up another avenue for the interpretation of the topological classification and the dynamical illustration of multilevel systems [217, 218]. It's also shown [97] that quantum geometry and the related quantum Fisher information could serve as a tool to characterize many-body correlations of the system, allowing us to identify strongly entangled phase transitions with a divergent multipartite entanglement.

Furthermore, we show that the quantum geometric tensor is closely connected with the precision bounds in quantum multi-parameter estimation problems. Specifically, for pure quantum states, the real part of QGT is proportional to the quantum Fisher information, while the imaginary part of QGT links the attainability of precision bounds when one estimates multi-parameters simultaneously. The quantumness of the system is represented by a characterization number $\gamma \in [0, 1]$ and is bounded by the fundamental uncertainty principle. With the previous experiment on measuring QGT of a three-parameter quantum state, we experimentally get this number and then extract the attainable quantum Cramér-Rao bound of the system.

Following the parameter estimation problem discussed above, we then switch to the practical sensing application side and discuss NV centers as quantum sensors to probe biological and chemical processes. The key idea of detecting biological or chemical signals is to transduce them into magnetic or electric noise that NV center is sensitive to. We first show that the longitudinal relaxation time T_1 of NV centers in nanodiamonds is sensitive to external magnetic molecules such as Gd complexes, which can be described by a simple theoretical model. Based on T_1 relaxometry, the rotational Brownian motion of magnetic molecules in solution is successfully probed. Furthermore, we design a highly-sensitive virus RNA sensor where the presence of RNA copies would trigger a distance change between NV centers and Gd magnetic molecules thus changing the T_1 relaxation time of NV centers. While we take the SARS-CoV-2 virus as an example, the proposal can be generalized into detecting other types of RNA virus or DNA virus. The proposed hybrid sensor is scalable, fast, and low-cost. The experimental demonstration of the virus sensor is undergoing while we are preparing theis thesis.

Indeed, due to the strong dipolar interactions between NV centers and Gd molecules and leveraging the NV center's excellent sensitivity in probing magnetic field, the NV-Gd pair has the potential to replace FRET-based sensors and be used in detecting more complicated biological or chemical processes. Compared to FRET molecules, nanodiamonds containing NV centers can have a better photo- and thermal- stability and better biocompatibility. With the assistance of chemical engineering, the target signal would trigger a distance change between NV centers and Gd complexes, i.e., change of magnetic environment of NV centers, which can then be probed by measuring the T_1 relaxation time in an all-optical manner or by measuring the resonance frequency change with the help of microwave pulses. We believe our work here will inspire more intriguing applications along these lines.

As the last part of the thesis, we design and demonstrate an alkali cation sensor, where the electrical noise from the cations on nanodiamond surface would sufficiently modulate the charge state of NV centers inside. The sensor is based on surface engineering of crown ether groups that can selectively interact with certain alkali ions. This gives another case where diamond surface chemistry technologies enable us to probe target signal that modifies the local environment of NV centers inside diamond.

Thanks to high photo-stability of NV centers, it's also an appealing direction to combine these nanodiamond-based sensors with state-of-art imaging technologies such as expansion microscopy [219, 220] to provide both superresolution and the capability of monitoring local environment changes dynamically. In this case, NV centers act as both fluorescent markers and highly-sensitive sensors and would provide new insights in biological research.